Energy, Climate and the Environment

The aim of this series is to provide texts which lay out the technical, environmental and political issues relating to proposed policies for responding to climate change. The focus is not primarily on the science of climate change, or on the technological detail, although there will be accounts of this, to aid assessment of the viability of various options. However, the main focus is the policy conflicts over which strategy to pursue. The series adopts a critical approach and attempts to identify flaws in emerging policies, propositions and assertions. In particular, it seeks to illuminate counter-intuitive assessments, conclusions and new perspectives. The intention is not simply to map the debates, but to explore their structure, their underlying assumptions and their limitations. The books in this series are incisive and authoritative sources of critical analysis and commentary, clearly indicating the divergent views that have emerged whilst also identifying the shortcomings of such views. The series does not simply provide an overview, but also offers policy prescriptions.

More information about this series at
http://www.palgrave.com/gp/series/14966

David Elliott

Renewable Energy in the UK

Past, Present and Future

palgrave
macmillan

David Elliott
The Open University
Milton Keynes, UK

Energy, Climate and the Environment
ISBN 978-3-030-04764-1 ISBN 978-3-030-04765-8 (eBook)
https://doi.org/10.1007/978-3-030-04765-8

Library of Congress Control Number: 2018963556

Cover illustration: Karl Johaentges/LOOK-foto/getty images

This Palgrave Macmillan imprint is published by the registered company Springer Nature Switzerland AG
The registered company address is: Gewerbestrasse 11, 6330 Cham, Switzerland

Acknowledgements and Web Links

Thanks are due to Tam Dougan for her advice and support. This book was her suggestion! Figure acknowledgements are included above with the Figures list. Parts of the text in the pre-2000 sections of this book use material from a 1997 publication of mine, 'Renewables: past, present and future', which in turn was based in part on a case study I produced in 1986 for the Open University course T362 Design and Innovation (Block 4). The 1997 report was for NATTA, the Network for Alternative Technology & Technology Assessment, which I helped set up in 1976. The bimonthly NATTA newsletter, later called Renew, which I started up in 1979, and which Tam Dougan, NATTA's long-serving co-ordinator, still works on, has also been a valuable source for this book. The free *Renew on-line* version, produced in parallel from 1996 onwards (starting at Renew No.100), offers a useful historical resource. Current and archived versions are at https://renewnatta.wordpress.com.

For the text references, I have provided web links wherever possible (inevitably more in the post-2000 period), all viewed in August 2018. I have tried to provide full references and a comprehensive and accurate account of events and views. I am grateful to several people who commented on drafts, including David Green and Jonathan Scurlock. My son, Oliver Elliott, also helped with advice on historical research. However, any errors are my own, and I will be glad to hear of any faults or failings and make amends as appropriate in any future editions.

Energy and Power Units

Strictly, energy cannot be generated or consumed, only converted from one form into another, with some losses, but it is common practice to talk of energy 'generating' or 'consuming' devices. They are given a **power rating** (or '**rated capacity**') in **watts** and multiples of watts, for example, a **kilowatt** ('kW') is 1000 watts. A **megawatt** ('MW') is 1000 kilowatts, a **gigawatt** ('GW') is 1000 MW and a **terawatt** ('TW') is 1000 GW. A typical large modern nuclear power station has a rated capacity of around 1 GW, while in the mid-1990s, the UK had around a total of 65 GW of electricity-generating capacity.

The amount of **energy** converted ('generated' or 'consumed') is defined by the power of the device multiplied by the time for which it runs (i.e. watts × hours). It is usually measured in **kilowatt hours** or '**kWh**'. This is the unit by which electricity, and gas, is sold. A typical 1 kW rated one bar domestic electric fire 'consumes' 1 kWh each hour. For larger quantities, multiples of kWhs are used, most commonly the **terawatt hour** (TWh) which is 1000,000,000 kWh. The national mid-1990s figure for total UK **electricity** 'consumption' was about 300 TWh per annum. Note that this is the figure for the consumption of *electricity*, not total *energy* consumption. Note also that the term 'power' is often used to mean electricity, that is, electric power, as in 'power plant'.

The **total** amount of energy used for all purposes is often measured in terms of **primary** energy consumption, that is, the amount of energy in the basic fuels used by energy conversion devices. In the UK in the past, given the historical preponderance of coal as a fuel, this was often measured in terms of the equivalent amount of coal that would be required to be burnt to provide that energy regardless of what fuel was actually used in power stations, that is, in 'tonnes of coal equivalent' (or, more, usually '**million tonnes of coal equivalent**' or 'mtce'). The mid-1970s figure for total UK primary energy was around 320 mtce p.a. For conversion, 1 mtce = 8141 GWh. 1 tonne is 1000 kilogrammes.

With the demise of coal and the advent of renewables like wind and solar, which do not use fuels as such, the mtce measure no longer has much relevance, but it has been used in the early parts of this study, since it was widely cited at the time. In later periods, when oil politics became more important, use was made of **million tonnes of oil equivalent** (mtoe), and that is still often used in modern scenarios. Clearly, a measure of tonnes of coal or oil inputs is helpful in thinking about climate impacts, but in modern energy terms, since renewables like wind and solar do not use fuel, what really matters is delivered energy, that is, what is available for final use. Note however that, confusingly, 'grossed up' primary energy figures for nuclear or renewables are sometimes cited, based on the amount of coal that would be needed to be fed to a coal plant to give the same electricity output.

Contents

About the Author

David Elliott is Emeritus Professor of Technology Policy at the Open University, UK. He has worked initially with the UK Atomic Energy Authority at Harwell and then for the Central Electricity Generating Board in Bristol, before moving, in the early 1970s, to the Open University, where he carried out research and developed courses on technological innovation, focusing in particular on renewable energy technology policy. Professor Elliott has written extensively on sustainable energy policy and is academic co-editor of Palgrave Macmillan's Energy, Climate and Environment book series.

Abbreviations and Chronology

Renewable Energy in the UK: Key Events Chronology

1970s	The Alternative Technology movement[s] starts to promote renewables
1973	Oil Crisis. Environmental lobby groups push for alternatives
1974	Election of a Labour Government, Establishment of ETSU
1979	Election of a Conservative Government
1982	ACORD recommends cut-backs on wave power. Wind lifts off in the USA and Denmark
1989	Privatisation of UK Electricity Supply Industry
1990	First NFFO set
1992	Bulk of Department of Energy absorbed into the DTI
1994	R&D cut-backs
1997	Election of Labour Government
1998	Market liberalisation begins
2000s	RO replaces NFFO, commercial interest grows. Small FiT introduced 2010s Conservative/coalition governments, CfD introduced, but cuts slow expansion

Support Scheme Abbreviations

CfD	Contracts for Difference
FiT	Feed-in Tariff
NFFO	Non-Fossil Fuel Obligation
RO	Renewables Obligation

Technology Abbreviations

CCS	Carbon capture and storage
CHP	Combined heat and power
PV	Solar photovoltaics

List of Figures

List of Tables

List of Boxes

1

Renewables: From the Fringe to Dominance

1.1 Introduction: A Surprising Success Story

Renewable energy sources are now supplying around 30% of UK electricity and their use seems set to continue to expand to supply maybe 50% by 2035, and possibly much more in the decades ahead (BEIS 2018). A study for the UK Governments National Infrastructure Commission looked at a scenario with 80% of electricity coming from renewables by 2050 (Aurora 2018). Indeed, in some ambitious scenarios, by 2050, renewables supply *all global power* and *all other energy needs* (Jacobson et al. 2017). That may be too optimistic, but there are more moderate scenarios in which renewables supply over 80% of global power by around 2050 (IRENA 2017). Fifty years ago, when the use of renewable energy sources was first being promoted by activists on the countercultural fringe, this sort of projection would have been almost inconceivable. At that point, hydro power and some academic work aside, all that existed, for example, in the UK, were a few amateur experiments, often with tiny home-made wind turbines and solar collectors, typically servicing rural retreats or 'alternative' community projects.

D. Elliott, *Renewable Energy in the UK*, Energy, Climate and the Environment,
https://doi.org/10.1007/978-3-030-04765-8_1

This book explores how that changed and how renewables became what now seem likely to be the dominant energy sources of the future. It focuses on the UK, although many of the developments it describes were related to, and went in parallel with, developments elsewhere, notably in the USA and Denmark, and more recently Germany, with the latter, along with China, now well ahead in the race to deploy renewables on a wide scale. The UK story is certainly not unique nor exemplary: as we shall see, many of the early initiatives were blocked or marginalised, and resistance to change has been a constant experience. But that has been overcome. This book shows how this came to be.

Inevitably, given that it is focused on major changes in policy and strategic approach, the emphasis is on the often-turgid details of institutional wranglings and contested funding programmes, rather than on the technology. The technical and strategic case for renewables is well explored in a range of texts, including in my own overviews (Elliott 2007, 2013), and more recently in some excellent updated editions of some well thought of standard text books (Peake 2017; Sorensen 2017). Instead, this book tracks through, in some detail, the policy and ideas battles from the early days to the present, and then offers a prognosis of what might happen next.

The story it tells is a familiar one in any area of technological innovation, of an uphill struggle against vested interests defending a well-established technological and institutional status quo. In 1981, one of the early renewable energy pioneers, Professor Stephen Salter at Edinburgh University, commented 'We are attempting to change a status quo which is buttressed by prodigious investment of money and power and professional reputations. For 100 years it has been easy to burn and pollute. 100 years of tradition cannot be swept away without a struggle. The nearer renewable energy technology gets to success, the harder that struggle becomes' (Salter 1981). He was right, but the battle seems to have been won, at least partly.

I have divided the story into two parts, before and after 2000, with an interim analysis in between. This 'end of century' division is somewhat pragmatic, but is not entirely arbitrary. The first period starts with the fringe pioneers and then looks at how the UK government (and other governments) took an increasing interest in renewables, stimulated by the

1973–1974 oil crisis. In the second period, from the year 2000 onwards, concerns about climate change, although already influential, became a major driver, along with increasing concerns about geopolitical security, especially following the terrorist attacks in the USA and elsewhere.

This approach highlights the point that the major drivers for change have been 'external' events, rather than the 'autonomous' emergence of new technologies (Winner 1977). The acceptance of the new energy technologies may have been aided by the work of the early pioneers and the lobbying activities of environmental groups, and the later successful take-up of the technologies was certainly aided by the efforts of innovators and engineers in helping the technologies to develop. However, that would probably not have happened if the wider context had not changed.

That said, this book does not explore the exact role that these and other external factors played. It is clear that the main 'external' events (the oil crisis and climate change) related back to energy use. So they could have been predicted (and indeed were by some). Even so, although they were not 'Acts of God', they came as a shock. The result was that there were changes, a reconfiguration of approach and an opportunity for new approaches to be adopted. This book focuses on these responses, and, in particular, on how renewable energy development was handled in the UK. It also inevitably touches on nuclear power, since that was, in effect, a rival non-fossil energy option. Initially it was the clear leader, with renewables as the challenger, but latterly its prospects have diminished somewhat, while renewables have gone from strength to strength. It should be noted that a key motivation for many of the early renewable energy supporters was to develop an alternative to nuclear, given its cost, safety, security and radioactive waste problems (Elliott 1978), and subsequently that motivation was reinforced by the major nuclear accidents in 1979 (Three Mile Island, USA) and 1986 (Chernobyl, USSR).

I should point out that I was an active participant, if only at times tangentially, in many of the debates over policy developments that are described and reviewed in this book, so it benefits from an 'insider' view. That also of course risks bias. While clearly supporting renewables, I have endeavoured to provide an impartial account of events and issues, although inevitably there will be those who see things somewhat differently. By focusing, in the chapters that follow, mainly on official reports,

I have tried to avoid a partisan approach, while also hopefully giving fair treatment to divergent views.

I have found few academic treatments of the overall story this book tells, but an excellent PhD thesis on the UK government's renewable energy programme by an academic historian covers the period 1974–1988 in a much more scholarly, detailed and archive-based fashion than I could achieve in my broader study (Wilson 2010). I was pleased to see that its conclusions were generally similar to my own, at least in terms of what happened up to 1988. It says 'the programme can be seen in some senses as a tokenistic gesture by the government acting within the uncertain political, social, and economic landscape of the 1970s'. However, as shown in this book, while that may well have been true initially, subsequently, despite continuing contestation and opposition, the situation changed.

One broader conclusion of the PhD thesis was that 'government decisions on renewable energy were continually driven by socio-political factors which overwhelmed the unreliable economic case for renewables at that time'. As we shall see, that tendency continued. Renewables were certainly treated as marginal for much of the initial period, but they did progress despite their problems, and, 30 years on, the economic uncertainty is now much less, with renewables becoming cost competitive and of much more central concern. Even so, I would recommend that any reader who needs a more detailed analysis of the early days takes a look at this work, which also includes helpful references to other early research studies in the field.

1.2 A Brief History of Post-War UK Energy Policy

Before plunging into the details, to set the scene, it is helpful to briefly summarise the post-Second World War energy situation in the UK, and how it initially developed. This short overview (and the timeline/chronology of key events offered earlier) obviously compresses a long history and a complex story, but there is no shortage of more detailed histories (Chick 2007), commentary (Pearson and Watson 2012) and contemporary analysis (Hall et al. 2016), and subsequent sections of this book go into exhaustive detail in terms of the development of renewable energy. However, the simple historical message is that the post-war context was not a promising one for new energy ideas, with a firmly established technological status quo blocking the way. It took some major challenges to change that.

Certainly, in the decade or so after the Second World War, the UK energy system seemed remarkably stable, based mainly on coal. There had been severe power shortages in the late 1940s, but the answer was seen, by the then Labour Government, as getting more coal and building more coal plants, with state-owned coal mines feeding the state-owned power system, and also supplying coal to make so-called town gas, to be distributed by the state-owned gas company for heating.

The 'forward ordering programme' for new coal plant construction, underpinned by the continued rise in demand for power, settled into what seemed to have become a standard '1–2 Gigawatts more' annual pattern, continuing like that well into the 1960s. The arrival of nuclear power on the scene at that point, using Magnox reactors, a technology spun off from the nuclear weapons' programme, did not change things much: the total nuclear energy input, initially, was small. The growth of oil as a major energy import was much more significant, but, until the first 'oil crisis' in 1973–1974, its availability was mostly seen as unproblematic. So the period of stability lasted for some while.

That is not to say there were not divergent views, even within the establishment. The National Coal Board was unhappy about the emergence of nuclear (a new fleet of Advanced Gas-cooled Reactors was being built during the 1970s and 1980s), which it saw as a threat. Meanwhile, outside of the establishment, the emerging environmental and conservation movements were also unhappy about the expansion of nuclear and also, more generally, about the overall growth-orientated direction being taken and its environmental implications.

However, it was the economic and geopolitical shock created by the 1973–1974 oil crisis that changed things. This was the result of an embargo imposed by OPEC, the then Arab-dominated Organisation of Petroleum Exporting Countries, targeted at nations perceived as supporting Israel during the Yom Kippur War, including the UK. It led to large, rapid global oil price rises (by up to four times) and significant shortages in many countries, including the UK, impacting on the public directly and visibly: vehicle fuels ran out and there were long queues at filling stations for what little was left. There were also major knock-on effects on the national (and global) economy, and consequent political instability, contributing, along with a major coal miners' strike and resultant three-day working week (to save energy), to the collapse of the then Conservative Government (under Edward Heath) in 1974. There was much talk of an 'energy crisis'. Some of that was linked to fears of impending resource exhaustion, with alarmist projections about oil reserves 'running out'. That might be some way off (it evidently still is), the more immediate issue at that time being price rises, and that, coupled with the issues of pollution and 'acid rain' production, started to make fossil fuels less reliable and attractive.

As a result, new ideas were now sought out. Although as we shall see, renewable energy ideas were explored in the UK, as elsewhere, the main expected technological beneficiary in the UK was nuclear. Ten years earlier, in 1963, Labour leader Harold Wilson had famously spoken about the 'white heat' of technological progress, and now, following the oil crisis, the new Labour Government, initially under Wilson, looked to a new nuclear programme. There was some debate as to what type of reactor to use, with a range of rival UK and US designs being on offer, but the government eventually opted for a UK design. However, that was not to be. Margaret Thatcher's Conservative Government, elected in 1979, opted for 10 US-type Pressurised Water Reactors, although, in the event, only one went ahead, Sizewell B in Suffolk. The Thatcher government also set about making some radical political and institution changes, notably, in the 1980s, the privatisation of the gas and then electricity industries and a cut-back of the once-dominant coal mining industry, after winning a final bitter struggle in 1984–1985 with the mine workers.

These political changes were controversial and arguably moved the UK 'from bad to worse' with a centralised state-led system replaced by one led by markets and the commercial interests of competing companies (Roberts et al. 1991). In institutional capacity terms, privatisation also meant the collapse of the major state-funded energy R&D base that had existed previously. Nevertheless, coupled with the search for new energy sources, privatisation did open the way for some technological changes, notably the use of gas for power generation. North Sea gas had been discovered much earlier and was being fed to the gas mains in the 1970s, replacing coal-derived town gas. But in the 1980s, following the privatisation of the electricity industry, and a change in policy on gas use, it was also used for power generation. The technological basis for a new period of quasi-stability had thus been established. There was a 'dash for gas', led by the newly privatised power companies, with the so-called natural gas used in cheap, efficient combined cycle gas turbines, while, although reliance on coal continued, it was reduced and mostly imported. So was oil, but North Sea oil provided a respite.

The 1973–1974 oil crisis had nevertheless been a major shock. It marked the starting point for most of the main subsequent structural and technological changes in the UK power system, as well as stimulating interest in alternative energy sources. It had also enabled environmental groups to challenge the overall direction of travel more aggressively, pointing to the limits to growth and the need for a new approach. Opposition to nuclear power remained a concern, but oil also became a major focus in the environmental agenda, aided by series of catastrophic oil spills around the world. In the late 1970s, the environmental movement also began to pick up on some of the ideas that has emerged on the fringe for alternative energy options. By the 1980s, these options were being taken more seriously and, as we shall see, with the climate change issue beginning to play more of a central role in energy policy considerations, the alternative sources moved to the fore. In effect, the second stable period had been disrupted by a new shock, climate change.

Although not linked to a specific event like the 1973–1974 oil embargo, awareness of the climate issue had an increasing and cumulative impact, and since they were, in the main, carbon free, the new set of alternative energy options, first looked at after the oil crisis, were even better placed to challenge the old set. Given that background, before we move on to look at what happened next, it is worth going back to the 1960s and 1970s, to see where the new 'Alternative Technology' ideas came from.

1.3 Alternative Technology: Ideas from the Fringe

As we have seen, powerful institutions had shaped the energy story in the UK as elsewhere, dominating the scene in the 1960s and early 1970s. However, they were not the main source of the new ideas; they mostly came from outside and often from the fringe. The 1960s saw the emergence of a range of radical movements, most of which sought to challenge what they saw as the deadening impact of the technocratic elite. Technology was obviously only one issue, but, given the range of environmental problems which were emerging from the technology of the day, it is not surprising that some radicals looked to alternative forms of technology that might avoid these problems. Its influence on the wider world can be overstated, but, arguably, key parts of the initial framing of the new approach to technology looked at in this book were the result of ideas from the so-called Alternative Technology (AT) movement, which emerged in the UK, USA and elsewhere in the late 1960s/early1970s, promoting, amongst other things, solar and wind power.

There had been earlier 'utopian' interest in such ideas in the USA (Herring 2005), but, a few experimental projects in the late 1800s and early 1900s apart (some are looked at below), it was mostly very general and abstract, whereas in the late 1960s, as the counter-culture and environmental movement spread, it became more practical, insistent and focused on the technology. Groups sprung up in the USA like the New Alchemists in New England, and the influential *Whole Earth Catalogue* from Stewart Brand in California brought together many ideas for a new approach to technology (Brand 1968).

The term 'Alternative Technology' (AT) was apparently first used in the UK in the early 1970s by Peter Harper, who became one of the editorial group that produced *Undercurrents*, the UK magazine of 'radical science and peoples' technology'. From 1972 onwards, it promoted radical alternatives to the existing range of technologies, as outlined in the seminal *Undercurrents* book *Radical Technology*, edited by Harper and *Undercurrents*' founder Godfrey Boyle, which consolidated views from the first few years of the magazine (Boyle and Harper 1976). I should add that I was also a regular contributor to *Undercurrents* from issue 10, until its demise in 1984. Back issues have been archived electronically (Undercurrents 2018) and offer a fascinating snapshot of views and issues. Its launching manifesto asked 'must technology remain no more than man's self-issued license to rape nature, rather than [becoming] a means whereby we can live in sympathy and harmony with the natural world?' (Undercurrents manifesto 1972).

In a parallel development, there was also growing interest in 'appropriate technology', emphasising products and systems for use in developing countries. This was promoted by the UK-based Intermediate Technology Development Group (ITDG), reflecting the ideas explored in the influential book *Small Is Beautiful* (Schumacher 1973). Their technical prescriptions had some overlaps with 'western' AT, though the developing country context was clearly different from that in industrial countries, where advanced technologies were well established and expanding rapidly.

At root, in the later, Western, context, the AT movement, although broad and sometimes a little chaotic, sought to promote socially and environmentally preferable *alternatives* to the existing range of technologies, emphasising simpler, human-scale options, often involving 'd-i-y' self-help 'backyard' approaches, and often being set in a community context. There was some congruence with the ideas popularised a century earlier by William Morris and the UK Arts and Crafts movement. In its early days, the AT movement certainly promoted craft-based artefacts and artisanship, and attempted to encompass almost all types of technology, including food production, transport and housing. However, the energy field tended to attract most attention, especially after the 1973–1974 oil crisis. The term 'Alternative Energy' became widely used, in the USA especially, denoting alternatives to fossil fuel and nuclear power, a set of new allegedly safer and cleaner energy options, including what are now called renewable energy technologies, one virtue of the latter being that they would not run out. In 1976, in an influential paper, and later a book, the US energy guru and Friends of the Earth activist Amory Lovins contrasted the 'soft energy path' using such technologies, with the 'hard' path using centralised fossil and nuclear technologies (Lovins 1977).

Clearly, the AT movement was international. In that context, mention should be made of the UN Summit on the Human Environment held in Stockholm in 1972, on the fringe of which an internationally oriented group called Powwow collected ideas for radical alternatives to the conventional approach being discussed in the main conference (Faramelli 1972). Initiatives like that helped spread the ideas internationally, with groups emerging around the world, for example, in Denmark, Germany and Australia, as well as the USA and the UK.

What emerged in the UK (the main focus of this book) was arguably more politically radical, reflecting its early overlap with the generally Leftish radical science and technology movement, as indicated by David Dickson's seminal 1974 book 'Alternative Technology: the politics of technical change' (Dickson 1974). It was also shaped by its countercultural roots and its subsequent links with the burgeoning environmental movement and the emerging anti-nuclear movement, even if the relationships between them were not always straight forward (Herring 2005). Similarly, there were sometimes disconnects and conflicts over the strategic approach with the political left (Elliott 2016) and with some of those behind the 'Blue Print for Survival' campaign and the Ecologist magazine (Goldsmith 1972). So there were some divergent political and strategic views. In this context, it would be remiss not to note that some environmentalists during this period were willing to condone the interim use of coal, suitably cleaned up, as a 'bridge to the solar future', although that later had to face the emerging issues of acid emissions, air pollution and then, crucially, climate change. It is also worth noting that some of the technologies promoted by the AT movement were adopted by the generally right-wing 'survivalist' groups in the USA, and live on today, as back-drops to some of the more lurid apocalyptic science fiction movies.

Back in the 1970s, however, the most immediate practical context for AT was provided by the various alternative community groups that had sprung up in the early 1970s (e.g. in the USA, Denmark and UK), aiming for a high degree of self-sufficiency. For that aim, energy was an obvious key requirement, so they pioneered the development of small wind and solar energy devices, along with biogas production and energy conservation measures. Alternative food production and housing options were also tested.

For some the aim was to create a personal or community lifestyle that felt right, rather than to change the wider world. Some though felt that, by developing convincing prototypes and 'living in the future now', they might act as catalysts for wider change. Over time, the emphasis shifted from isolated niches in 'retreatist' communities to a more outward-looking approach. Some of the communities, like the long-running Centre for Alternative Technology (CAT) in rural Wales, which was set up in 1972–1973, served as working exemplars or demonstration centres, not least so as to earn income from visitors. A multifaceted account of CAT's early days has been archived electronically and provides some fascinating insights into the changing experiences there (CAT 1995). In parallel with rural projects like CAT, some AT activists focused on urban projects, fleshing out the idea of community-led alternatives, so-called Community Technology. A core belief in much of this was that physical demonstrations were vital. As pioneering green architects Brenda and Robert Vale put it: 'One live, working experiment, however impractical if it were applied universally, will transmit an idea far better than a shelf full of theoretical reports. Something that can be seen and touched and shown to work to some degree arouses curiosity, and curiosity in turn leads to solutions' (Vale and Vale 1975).

The labels used by the various groups differed (e.g., the Northumbria-based group was initially called 'New Age Access'), but they all, in their different ways and locations, sought to demonstrate practical examples of what could be done to develop what one group called 'Rational Technology', as in the Rational Technology Unit run by wind energy pioneer Derek Taylor and others based at the Architectural Association in London. Many of these groups came together at 'Comtek', an annual AT festival in Bath.

However, it was often not easy to be taken seriously in the wider word. Detractors could portray the new often self-built technologies as irrelevant and unworkable in terms of modern energy needs and lifestyles, a romantic retreat to the inefficient energy systems of the pre-industrial past. That may ignore the fact that the early stages of the industrial revolution were in fact powered by water mills and that hydro now supplies about 18% of the world's electricity. Nevertheless, it was easy, given the glamour and apparent success of modern high technology, for critics to denigrate the d-i-y level efforts of the AT enthusiasts and also to ridicule their counter-cultural values: crazy ideas, thrown up by a bunch of crazy hippies. Although the reality was usually a little more complex: see Box 1.1.

Box 1.1 CAT: Ideology in Action?

There was certainly idealism and naivety in the AT movement, but also a sense of purpose, as the 1995 retrospective CAT review makes clear. Gerrard Morgan-Grenville, an old Etonian and businessman-turned environmentalist, who set up what was initially called the National Centre for Alternative Technology in Wales in 1972, reported how, just prior to that, he had taken a sabbatical and gone to America. 'I studied groups who were trying alternatives, mostly hippies. Many of them were attempting some of the technologies that C.A.T. went on to demonstrate, but not very successfully. They were mostly muddled and disorganised, fragile. Most failed, but the important thing was that they had opted out of mainstream life in order to find a way of living which respected the environment in which they lived. They also rejected Authority in principle. I found myself in sympathy with both aims' (CAT 1995).

Life at CAT in the early days was not exactly utopian: it was a rain-swept abandoned quarry. But it clearly attracted people with skills as well as ideology. Tim Kirby, who became CAT's site engineer, said 'It felt just exactly what I wanted to do, both to use what I'd learnt in a productive way and live the way I wanted to live. It was a brilliant mixture, from unblocking sewers, to replacing rat-chewed wiring, to teaching fairly advanced technical things on a wind-power course, or water... It was about a quartering of salary, but I didn't notice that. I felt richer than I've ever felt' (CAT 1995).

So it was something of an adventure that attracted a range of people for a range of reasons, with CAT striving to be self-sufficient in energy and other needs. However, the focus changed over the years. In the 1995 review, CAT development director Paul Allen said 'Self sufficiency is dead, not because it's not green – but because it's not replicable! We now need to explore the most effective means of reducing our environmental footprints. Dropping out to build a better world has given way to the challenge of developing ideas, tools and technologies which will inspire, inform and enable mainstream society to join the process of sustainable development' (CAT 1995).

The energy technologies that were promoted by the early AT enthusiast were certainly unusual, radically different technologies. Moreover, the term 'radical' also means 'going back to the roots', and many of the ideas that were promoted were indeed often not new. Many of them had been explored earlier, some of them quite extensively. Wind power has of course been used around the world since antiquity, and, for example, in the Middle Ages for corn grinding. More recently, small multi-vane wind-driven water pumps were very widely used in the USA and Australia, and, in the 1930s, some windmills were used to generate electricity on farms in the USA, where distribution systems had not yet been installed. Small wind-electric chargers were also extensively used in Denmark. Some later moved up scale. An innovative 200 kW wind turbine was built in 1956–1957 by Johannes Juul for the electricity company SEAS at Gedser in Denmark. But the largest project from the early days was in the USA. A giant 1.25 Megawatt wind turbine was built in 1941 in Vermont by local enthusiasts, with some engineering company help, at a site known as Grandpa's Knob (Putnam 1948).

The use of solar energy also has a very long history, from ancient Greece onwards, with a wide variety of solar heat-using systems being developed over the years. In more recent times, solar heat collectors for water heating were widely used in California and Florida in the 1900s (Butti and Perlin 1980). The use of water power also has a very long history for corn grinding, and small water mills have been used for power production, as of course are large hydro plants. There were also early examples of biogas production from animal waste.

Interest in ideas for novel energy system has always fired the enthusiasm of lone inventors and innovators, as well as some academics, so there was a relatively rich, if sometimes undeveloped, field for exploration by AT enthusiasts. Certainly some of the early devices and projects reported in the literature, going back many years, were quite astounding: see Box 1.2. In the UK context, there were also some more recent examples to explore, like the 100 kW wind machine built on the Orkneys in 1955 (Ford 2018).

Box 1.2 Nothing New Under the Sun

In the 1870s, Augustin Mouchot, a professor of mathematics at the Lycee de Tours, used a conical reflector to focus solar heat so as to raise steam to drive a steam engine, and he even developed a system to generate electricity using solar heat and the thermoelectric effect. His assistant, Abel Pifre, went on to exhibit a solar-powered printing press at the Gardens of the Tuileries in Paris in 1880. Similar, but larger-scale, steam raising systems were then developed in the USA, by, amongst others, Aubrey Eneas, who built a focused solar water pumping system for farm irrigation. Reputedly, it could pump 1400 gallons of water per minute. A much-improved variant, with parabolic trough reflectors, was subsequently developed in the USA by Frank Shuman, and in 1912 a version was installed in Egypt, at Meadi, 15 miles from Cairo. It could pump 6000 gallons of water per minute and run 24 hours a day, since a large hot water tank was attached, storing some of the day-time heat (Butti and Perlin 1980). Hundred years or so later, that idea is now back in vogue, although for electricity production, with large concentrated solar power plants, backed up by molten salt heat stores, now running in North Africa and the Middle East, as well as Spain and the USA.

As noted earlier, wind-driven water pumps have had a long history in the USA, and versions were adapted there for electricity generation, but the first wind turbine used for the production of electricity was actually built in Scotland in July 1887 by Professor James Blyth of Anderson's College, Glasgow (the precursor of Strathclyde University). Blyth's 10-meter-high, cloth-sailed turbine was installed in the garden of his holiday cottage at Marykirk in Kincardineshire and was used to charge accumulators to power the lighting in the cottage. Blyth offered the surplus electricity to the people of Marykirk for lighting the main street, but they turned it down as they thought electricity was 'the work of the devil'. An even more advanced system was developed in 1891 by Danish scientist, Poul la Cour, who built a wind turbine to generate electricity which was then used to produce hydrogen by electrolysis, the gas being stored for use in experiments and to light the nearby Askov High school (Price 2005). One hundred twenty years on and we are now seeing that idea being fully developed in modern wind turbine-driven 'Power to Gas' projects in Germany and elsewhere.

The modern AT enthusiasts not only revisited and revived some of the idea from the past, but also took on board some of the more recent developments, such as photovoltaic solar cells, which were then being developed for use with space satellite and communications systems. An eclectic mix! The AT enthusiasts, and the small band of academics also active in the field, made some progress with developing some of these ideas, but, in general, the main contribution of the AT activists was not engineering, it was in promoting the idea that there were alternatives, and also offering a vision of how technologies like this, if fully developed and widely adopted, could form the basis for a new type of social order, an alternative society. Indeed, maybe optimistically, AT activist Peter Harper claimed that 'premature attempts to create alternative social, economic and technical organisation for production can contribute in a significant way to the achievement of political conditions that will finally allow them to be fully implemented' (Harper 1974).

This type of utopian vision may not have been realised outside of a few experimental communities, but the AT movement did begin to put the technologies on the wider agenda, if nothing else by providing positive technological options, and policy recommendations based on them, for environmental campaigners to point to. In 1977, CAT in Wales produced a pioneering report outlining how renewables could expand to supply the bulk of UK power (Todd and Altry 1977). Moreover, sometimes, in some locations, it was possible to go beyond just offering small projects or theoretical options, as in the case of the 2 MW Tvind wind turbine built in 1975–1978 by senior school students in Denmark. It was for a long time the largest wind machine in the world. It helped create the momentum that led local Danish agricultural engineers to come up with new (though smaller) turbine designs for use on farms, and produce what turned out to be very successful wind devices, expanding in size to become commercial scale (Karnoe 1990).

That is how Vestas and Bonus started out as small wind turbine manufacturing companies. They gradually expanded to become world-beaters, for a while dominating the world market as wind power took off in the 1980s. Within Denmark, they supplied wind turbines to the many new local wind co-ops that grew up across the country, with around 80% of the wind projects being locally owned by farmers and wind co-ops, aided by a Feed-in Tariff (FiT) support system.

That sort of transfer, from the fringe to entrepreneurial lift-off and market uptake, did not occur significantly in the UK. It had been aided in Denmark by the long and successful grass-roots environmental campaign against nuclear power, culminating in a national vote in 1985 against it. That referendum created the space for wind power, already developing fast, to expand further. It was similar, later on, in Germany, with a powerful anti-nuclear movement, also strongly committed to renewables as the alternative.

The situation was very different in the UK, where, despite attempts by the generally much weaker anti-nuclear and green movements, the nuclear lobby remained strong. The political and economic climate also made it much harder for local renewable energy projects to start up and for renewables to be taken seriously. However, in the wake of the 1973–1974 oil crisis, as noted above, most western governments started to look for alternatives, and that included the UK government. Although, as we have seen, the UK government looked mainly to nuclear, and gas began to play a key role, as renewables began to be taken up elsewhere, for example, in Denmark (as above) and also in the USA (under President Carter's Solar programme), the UK government also started to look at renewables.

Clearly, the main reason for that was a major 'external' event—the oil shock. But the lobbying work that had been done by environmental groups and the input from the Alternative Technology movement no doubt had some influence, talking up the alternative options. However, as we shall see, when it came to look at renewables, initially from 1974 onwards, and then more seriously in the 1980s and beyond, the UK government's approach was to take developments off in a somewhat different direction to what the AT movement had envisaged.

In tracking through the subsequent UK renewable energy story in some detail in the chapters that follow, one of my aims has been to assess the extent of the impact of AT ideas. As we shall see, renewable technologies *were* adopted, if slowly at first, but this might have happened anyway. Nevertheless, it does seem likely that the pioneering efforts of the AT movement played some role. If so, then this type of grass-roots innovation may have wider implications for technological development generally (Smith and Stirling 2018).

Modern-day innovation theorists talk in terms of the transfer of ideas from innovative 'niches' to the wider economy, with 'green' niches being one focus. For example, Smith sees these niches as 'spaces where networks of actors experiment with, and mutually adapt, greener organizational forms and eco-friendly technologies', and he says 'analysis of these greener 'socio-technical configurations' has inspired normative claims for niches as significant sites of learning and network building relevant for sustainable technology policy more widely' (Smith 2007).

Looking at the transfer process, he notes that 'unsurprisingly, green niches are more likely to diffuse into the mainstream, and thereby displace incumbent "socio-technical regimes", if the latter are placed under concerted pressure to become more sustainable'. That is very much the story told here, although as we shall see, to the extent that there was a 'translation' process, it led to some significant modifications in what was being transferred. As Smith puts it 'technologies favoured by niche actors, or user relations essential to niche performance, will be interpreted under the very different circumstances of the regime and considered in comparison with existing technological practices, or skills attributes, or market base'.

As we shall see, that is mostly what happened in the case of renewables and AT in the UK. To the extent that anything was transferred, either directly or via some of the AT-inspired educational courses that sprung up in some Universities, what emerged was often somewhat different from what many in the AT movement had in mind. That possibility was certainly something many AT enthusiasts had worried about. They did not relish their ideas being co-opted, shorn of the linked social values, and possibly corrupted. As Smith makes clear in a retrospective review, that may be the price of influence, or at least one of the risks (Smith 2005). However, the positive pay-off might be a change in the way things were done in the wider world, even if only marginally. Did the AT enthusiasts succeed in that? The study that follows may offer some clues, as we track though the way the UK government handled the development of renewable energy.

This historical review also offers some more general insights into how to manage the development of new technologies, although, it might be said, with some notable exceptions, mainly by providing examples of how not to do it! That is probably too cynical a view. While errors were certainly made, and much of the AT ethos abandoned, the UK has nevertheless managed to start making the transition to renewable energy.

References

Aurora. (2018). *Power Sector Modelling: System Cost Impact of Renewables*. Report for the National Infrastructure Commission, Aurora Energy Research. https://www.nic.org.uk/wp-content/uploads/Power-sector-modelling-final-report-1-Aurora-Energy-Research.pdf

BEIS. (2018). *Updated Energy and Emissions Projections 2017*. London: Department for Business Energy and Industrial Strategy. http://www.gov.uk/government/uploads/system/uploads/attachment_data/file/671187/Updated_energy_and_emissions_projections_2017.pdf

Boyle, G., & Harper, P. (1976). *Radical Technology*. London: Wildwood House.

Brand, S. (1968). Whole Earth Catalogue. *Multiple Subsequent Self-published Issues.*

Butti, K., & Perlin, J. (1980). *A Golden Thread: 2500 Years of Solar Architecture and Technology*. Palo Alto: Chelsea Books.

CAT. (1995). *Crazy Idealists.* Centre for Alternative Technology 25 Years Archive. http://archive.cat.org.uk/images/document_archive/Crazy_idealists.pdf

Chick, M. (2007). *Electricity and Energy Policy in Britain, France and the United States Since 1945*. Cheltenham: Edward Elgar.

Dickson, D. (1974). *Alternative Technology: The Politics of Technical Change.* London: Fontana.

Elliott, D. (1978). *The Politics of Nuclear Power*. London: Pluto Press.

Elliott, D. (Ed.). (2007). *Sustainable Energy: Opportunities and Limitations.* Basingstoke: Palgrave. Online at https://link.springer.com/book/10.1057%2F9780230378384

Elliott, D. (2013). *Renewables: A Review of Sustainable Energy Supply Options.* Bristol: Institute of Physics Publications.

Elliott, D. (2016). The Alternative Technology Movement. *Science as Culture, 25*(3), 386–399.

Faramelli, N. J. (1972). Toying with the Environment and the Poor. A Report on the Stockholm Environmental Conference. *Boston College Environmental Affairs Law Review, 2*(3), 469–486.

Ford, J. (2018, July 17). July 1955: The UK's First Grid-Connected Wind Turbine. *The Engineer.* https://www.theengineer.co.uk/orkney-wind-turbines/

Goldsmith, E. (1972). *Blue Print for Survival.* The Ecologist Special Report, London.

Hall, J., Tran, M., Hickford, A., & Nicholls, R. (Eds.). (2016). *The Future of National Infrastructure.* Appendix A 'An Overview of the UK Energy System Evolution in a Global Context, Current State and Future Outlook'. Cambridge University Press. https://itrc.ouce.ox.ac.uk/wp-content/PDFs/AppendixA-Energy-system-overview.pdf

Harper, P. (1974). What's Left of AT. *Undercurrents, 6*, 35–38. London

Herring, H. (2005). *From Energy Dreams to Nuclear Nightmares.* Oxford: John Carpenter.

IRENA. (2017). *Perspectives for the Energy Transition: Investment Needs for a Low-Carbon Energy System.* International Renewable Energy Agency, Abu Dhabi, Joint Report with the International Energy Agency, Paris. http://www.irena.org/menu/index.aspx?mnu=Subcat&PriMenuID=36&CatID=141&SubcatID=3828

Jacobson, M., Delucchi, M., Bauer, Z., Goodman, S., Chapman, W., Cameron, M., Bozonnat, C., Chobadi, L., Clonts, H., Enevoldsen, P., Erwin, J., Fobi, S., Goldstrom, O., Hennessy, E., Liu, J., Lo, J., Meyer, C., Morris, S., & Yachanin, A. (2017). 100% Clean and Renewable Wind, Water, and Sunlight All-Sector Energy Roadmaps for 139 Countries of the World. *Joule, 1*(1), 108–121. http://www.sciencedirect.com/science/article/pii/S2542435117300120

Karnoe, P. (1990). Technological Innovation and Industrial Organisation in the Danish Wind Industry. *Entrepreneurship & Regional Development, 2*, 105–123.

Lovins, A. (1977). *Soft Energy Paths*. London: Penguin.

Peake, S. (ed). (2017). *Renewable Energy: Power for a Sustainable Future* (4th ed.). Oxford: Oxford University Press. (Earlier Editions ed. Boyle. G.)

Pearson, P., & Watson, J. (2012). *UK Energy Policy 1980–2010: A History and Lessons to be Learnt*. Institution of Engineering and Technology/Parliamentary Group for Energy Studies. http://sro.sussex.ac.uk/38852/1/uk-energy-policy.pdf

Price, T. (2005). James Blyth – Britain's First Modern Wind Power Pioneer. *Wind Power Engineering*. http://journals.sagepub.com/doi/10.1260/030952405774354921

Putnam, P. (1948). *Power From the Wind*. New York: Van Nostrand Rheinhold.

Roberts, J., Elliott, D., & Houghton, T. (1991). *Privatising Electricity: The Politics of Power*. London: Belhaven.

Salter, S. (1981, August). Wave Energy: Problems and Solutions. *Journal of the Royal Society of Arts Proceedings*, 580. London.

Schumacher, E. F. (1973). *Small Is Beautiful* (1980th ed.). London: Blond and Briggs.

Smith, A. (2005). The Alternative Technology Movement: An Analysis of Its Framing and Negotiation of Technology Development. *Research in Human Ecology, 12*(2), 106–119.

Smith, A. (2007). Translating Sustainabilities Between Green Niches and Socio-Technical Regimes. *Technology Analysis & Strategic Management, 19*(4), 427–450.

Smith, A., & Stirling, A. (2018). Innovation, Sustainability and Democracy: An Analysis of Grassroots Contributions. *Journal of Self-Governance and Management Economics, 6*(1), 64–97.

Sorensen, B. (2017). *Renewable Energy* (5th ed.). London: Academic Press-Elsevier.

Todd, R., & Altry, C. (Eds.). (1977). *An Alternative Energy Strategy for the UK*. Machynlleth: Centre for Alternative Technology.

Undercurrents. (2018). Electronic Archive of All Back Issues: https://undercurrents1972.wordpress.com

Undercurrents Manifesto. (1972). https://issuu.com/undercurrents1972/docs/uc_manifesto

Vale, B., & Vale, R. (1975). *The Autonomous House*. London: Thames and Hudson.

Wilson, J. C. (2010). *A History of the UK Renewable Energy Programme, 1974–88: Some Social, Political, and Economic Aspects* (PhD Thesis). University of Glasgow. http://theses.gla.ac.uk/3121/1/2010WilsonJohnPhD.pdf

Winner, L. (1977). *Autonomous Technology: Technics-out-of-Control as a Theme in Political Thought*. Cambridge MA: MIT Press.

2

The Government Takes an Interest *in the 1970s*

2.1 The UK Government's Involvement

The UK government started taking an interest in renewable energy in the early 1970s, and in the chapters that follow, I track through the story from then to the present day, starting in this chapter with the early exploratory phase in the 1970s.

The full story of the government's handling of renewables is quite complex, with many government departments changing names and Ministers and governments coming and going. To aid navigation, in addition to the overall timeline/chronology of key events presented earlier, Box 2.1 provides a detailed breakdown of key Ministers, Departments, Agencies and governments covering the whole period.

D. Elliott, *Renewable Energy in the UK*, Energy, Climate and the Environment,
https://doi.org/10.1007/978-3-030-04765-8_2

Box 2.1 A Guide to Key Government Actors in the Story 1974–2018

UK Governments		Prime Ministers
1974–1979	Labour	Harold Wilson (1974–1976) James Callaghan (1976–1979)
1979–1983	Conservative	Margaret Thatcher
1983–1987	Conservative	Margaret Thatcher
1987–1992	Conservative	Margaret Thatcher (1987–1990) John Major (1990–1992)
1992–1997	Conservative	John Major
1997–2001	Labour	Tony Blair
2001–2005	Labour	Tony Blair
2005–2010	Labour	Tony Blair (2005–2007) Gordon Brown (2007–2010)
2010–2015	Con-Lib Dem	David Cameron (Coalition)
2015–2017	Conservative	David Cameron (2015–2016) Theresa May (2016–2017)
2017–	Conservative	Theresa May

Key UK Government Departments and Agencies

BEIS	Department for Business, Energy and Industrial Strategy (2016–ongoing)
BERR	Department for Business, Enterprise & Regulatory Reform (2007–2009)
BIS	Department for Business, Innovation and Skills (2009–2016)
DECC	Department of Energy and Climate Change (2008–2016), then absorbed in BEIS
DEn	Department of Energy (1974–1992), then the DTI took most of its work back
DIUS	Department for Innovation, Universities and Skills (2007–2009), then absorbed in BIS
DTI	Department of Trade and Industry (1970–2007), energy moved to DEn in 1974–1992
ETSU	Energy Technology Support Unit (1974–1996), then became part of AEA Technology plc
OST	Office of Science and Technology (latterly Innovation) (1992–2007)
PEnI	The DTI was briefly (for a week in 2005) renamed the Department for Productivity, Energy and Industry

Key Government Senior Figures in Energy

Tony Benn (Labour) Secretary of State for Energy (DEn) 1975–1979

David Howell (Con) Secretary of State for Energy (DEn) 1979–1981

Nigel Lawson (Con) Secretary of State for Energy (DEn) 1981–1983

Peter Walker (Con) Secretary of State for Energy (DEn) 1983–1987

Cecil Parkinson (Con) Secretary of State for Energy (DEn) 1987–1989

John Wakeham (Con) Secretary of State for Energy (DEn) 1989–1992

Within DTI Tim Eggar (Con) Energy Minister (DTI) 1992–1996

John Battle (Lab) Energy Minister (DTI) 1997–1999

Helen Liddell (Lab) Energy Minister (DTI) 1999–2001

Peter Hain (Lab) Energy Minister (DTI) Jan–Jun 2001

Brian Wilson (Lab) Energy Minister (DTI) 2001–2003

Stephen Timms (Lab) Energy Minister (DTI) 2003–2004

Mike O'Brien (Lab) Energy Minister (DTI) 2004–2008

Ed Miliband (Lab) Secretary of State (DECC) 2008–2010

Chris Huhne (Lib Dem) Secretary of State (DECC) 2010–2012

Ed Davey (Lib Dem) Secretary of State (DECC) 2012–2015

Amber Rudd (Con) Secretary of State (DECC) 2015–2016

Greg Clark (Con) Secretary of State (BEIS) 2016–ongoing

Some Key Energy Ministers

Eric Varley (Lab) DEn 1974–1975

Alex Eadie (Lab) DEn 1974–1979

John Moore (Con) DEn 1979–1983

David Mellor (Con) DEn 1981–1983

David Hunt (Con) DEn 1984–1987

Michael Spicer (Con) DEn 1987–1989

Colin Moynihan (Con) DEn 1990–1992

Key **DTI** energy ministers shown left (DTI Secretaries of State not shown)

DTI Civil servant Directors of its New and Renewable Energy programme:

– Godfrey Bevan (until 2001)

– John Doddrell (from 2011)

Malcolm Wicks (Lab) DTI/BERR 2005–2008

Charles Hendry (Con) DECC 2010–2012

Greg Barker (Con) DECC 2010–2015

John Hayes (Con) DECC 2012–2013

Andrea Leadsom (Con) DECC 2015–2016

Claire Perry (Con) BEIS 2017–ongoing

Chief Scientists: Government Chief Scientific Advisors
Some key CSA figures attached to Energy or related departments:

Walter Marshall	(DEn)	1974–1977	Head of the UKAEA (from 1981) and then (1983) the CEGB
Herman Bondi	(DEn)	1977–1980	
Anthony Challis	(DEn)	1980–1983	
Derek Pooley	(DEn)	1984–1987	
Robert May	(Government Chief Scientist)	1995–2000	
David King	(Government Chief Scientist)	2000–2008	
David MacKay	(DECC)	2009–2014	
John Loughhead	(DECC/BEIS)	2014–in post	

2.2 The Exploration Phase: ETSU Takes a Lead

Although there had been some interests in ideas like tidal barrages in the past, and some work on wind energy following the Suez crisis in the 1950s, the UK government's first significant involvement with renewables was in response to the 1973–1974 oil crisis. In January 1974, the Conservative Government set up a Department of Energy (DEn), and this new body was backed by the incoming Labour administration, after the fall of the Heath government in February. The DEn was represented in Cabinet by the Secretary of State for Energy, with a number of Ministers, with one or more responsible for renewable energy.

In April 1974, the Department set up an Energy Technology Support Unit (ETSU), based, a little oddly some thought, within the UK Atomic Energy Authority (AEA) research labs at Harwell. The idea of 'diversification' at Harwell had already been discussed by, amongst others, Lord Rothschild, chair of the Central Policy Review Staff (Wilson 2010), but the oil crisis gave that a new sense of urgency and purpose. Harwell's Director, Dr Walter Marshall, oversaw ETSU's creation. He was also appointed as the DEn's Chief Scientific Advisor. Since the Chief Scientist's support office was also based at Harwell, it made sense for ETSU to be there too.

ETSU, headed by Dr Keith Dawson, was charged with, amongst other things, providing technical advice, strategic assessment and management support for the Department's renewable energy programme, with a series of technology programme sub-groups and steering committees set up (Flood 1986). In most cases, the emphasis was on 'desk' studies, but some hardware projects were also initiated and carried out under contract, for example, by engineering consultants, research institutes and University research teams: see Box 2.2.

Box 2.2 The Initial R&D Programme

The primary aim of the first stage of the UK renewable energy programme was to map out overall possibilities and to see which technologies were most appropriate for further work. Some of this involved practical projects. It was engineering led, with **wave energy** emerging as an initial front runner, which, in his detailed review, Wilson attributes in part to the early influence of Lord Rothschild (Wilson 2010). In the wave-power field, several teams were funded separately, during the first phase, developing rival systems, the Salter duck, the Cockerell raft, and so on joined later by some new devices like the 'Airbag' and 'Clam'. Open water trials of scale models were carried out by the usually University-based teams for some of these systems (on the Solent and on Loch Ness).

In the first phase of the **wind-power** programme, design studies and assessments were the main focus, together with work on small prototypes, such as the Musgrove H-shaped vertical axis machine. **Solar** work focused mainly on providing grants to local authorities and monitoring contracts for universities for solar heating work and some solar photovoltaics (PV) work. For the other sources, there were desk studies, like that on the Severn **Tidal Barrage** carried out by the Bondi Committee, tidal power being seen as potentially large source. **Geothermal** energy was also looked at (both deep drilling for power and aquifer use for heat), as was **hydro** power and **biomass**, mainly, at this stage, by desk studies.

The main focus was on novel engineering options, and, as the programme unfolded, the emphasis was mostly on the large scale, the establishment view evidently being that only large-scale plants would be viable. So the research and assessment parameters included a 2 Gigawatt (GW) reference design for the wave energy programme, large multi-GW scale tidal barrages and large multi-MW scale wind turbines, although, at least initially, the later was not seen as having high priority. Solar too was not seen as a very major option. So the initial perspective was very different from that promoted by the Alternative Technology movement, which had emphasised small-scale local projects.

The difference in approach was perhaps inevitable, given the very different political perspectives and the hard engineering approach adopted by ETSU. Clearly, with little existing established research to draw on, ETSU faced difficult problems in trying to assess the viability of the various renewables. There were also inevitably limited funds and little direct experience with the novel sources and associated engineering requirements, and a wide variety of interest groups and inventors keen to promote their own ideas. Nevertheless, within the limits of an establishment-defined programme, ETSU did some excellent desk studies and assessments, acting essentially in an advisory role to government, as well as managing those projects deemed worthy of funding.

ETSU produced what became a series of 'overview' papers, which included reviews of the long-term potential of renewables, along with reports on specific technologies, published in the Department of Energy's 'Energy Paper' (EP) series. For example, Energy Paper 11, *Energy Research and Development in the UK*, was produced in 1976 for ACORD, the Governments Advisory Council on R&D. This included the estimate that the long-term (2025) potential for wave power might be as high as 50 million tonnes of coal equivalent (mtce) per annum, which it noted was 'about one half of the UK's requirement for electricity'. (See the note in the introduction on energy units for an explanation of mtce.)

Even though the overall conclusions concerning the potential role of renewables were fairly cautious, official estimates like this, and the ones which followed, coupled with more optimistic estimates emerging from external experts and bodies, and reports of progress overseas (e.g. in particular the USA), fired enthusiasm for renewables, at least in some circles. In general, the non-official view amongst *external* energy experts certainly was optimistic, especially about the longer-term future:

Solar: The UK section of the International Solar Energy Society (a highly respectable body of scientists and engineers) estimated that by 2020 Britain could be obtaining about 10% of its primary energy requirement (35 mtce) from solar sources (UK-ISES 1976).

Wind: Dr Peter Musgrove, a leading wind-power researcher, estimated that by 2025 50% of Britain's electricity could come from (mainly) offshore windmill units. He also suggested that 6% of the country's primary energy could be obtained from water turbines in streams and rivers (CAT 1977).

Wave: Professor Peter Chapman of the Open University Energy Research Group estimated that by 2025 the UK could obtain up to 120 mtce from wave-power systems (Chapman 1977).

There were also inputs on wind and heat storage from Sir Martin Ryle, the Astronomer Royal, who had become a supporter of renewable energy (Ryle 1977).

In 1977, the House of Commons Select Committee on Science and Technology produced its own report, having taken evidence from a large number of external as well as government experts. Its conclusions were very optimistic. The Committee was impressed by the potential contributions that might be available from solar (15 mtce by 2025), on-land wind (7 mtce) and wave (50 mtce) and also pointed to the considerable *offshore* wind potential (10–12 mtce): 'We recommend that the investment programme in R & D in renewable sources of energy should be expanded so that those renewable sources which prove to be technically and economically viable are in a position to begin making a worthwhile contribution to the United Kingdom energy requirements by 1990 so that when self-sufficiency in indigenous fossil fuel energy sources is past, they are well established. This target will require an increase in investment in research into renewable sources but we doubt if in the near future a sum exceeding £50 million per annum will be needed.'

It went on: 'We have no hesitation in recommending increased expenditure in the present climate of economic restraint since we are of the opinion that it will be fully justified by the medium and long term benefits. We believe that the funds which will be employed in the development of renewable sources will prove to be modest in comparison with the funding required by the nuclear programme, and by the coal mining, oil and gas industries over the years' (Select Committee on Science and Technology 1977).

The government's response, in its 1978 White Paper on Alternative Energy Sources, although somewhat more measured, was still basically positive: 'The Government [...] agrees with the Committee that the development of these alternative energy sources should be pursued with urgency and determination and that investment in research and development on such sources should be expanded. It has decided, in the light of progress and results so far, to commit a further £4.5 million to research and development.' However, it added that it did 'not consider it practicable at the present time to fix a target date by which the renewable sources should be making a significant contribution' (HMG White Paper 1978).

Some details of the Government's assessment of the technologies at this point are given in Box 2.3 from the White Paper. Note the low priority given to wind power and solar, as compared with wave and tidal power.

Box 2.3 The 1978 White Paper

The White Paper on alternative sources represented the [Labour] Government's reply to the Third and Fourth Reports from the Select Committee on Science and Technology, Session 1976–77. It made the following observations on the potential contributions from alternative sources:

> *Solar Energy* 'The Government agrees with the Committee that solar water heating has the most immediate potential of all the renewable sources in that systems for the pre-heating of domestic and industrial water supply are already commercially available. But the capital cost of such systems is in general high, in relation to the savings of conventional fuel that can be achieved through their use.'
>
> *Wind Power* 'The Committee saw little present potential for the use of wind power for electricity generation. The assessment prepared by ETSU (Energy Technology Support Unit) identified a limited scope for the generation of electricity on most favoured hilltop sites, at costs possibly competitive with fossil fuels.'

Wave Power 'Wave Power constitutes a major potential, as the Committee concluded, for large-scale electricity generation, provided the considerable technical problems can be overcome and economic viability can be established. Progress in the wave power programme which was launched in 1976 has been encouraging. Of the devices under laboratory examination, work on two has already advanced from small-scale laboratory tests to trials at one-tenth scale in open water, on Loch Ness and Solent.'

Tidal Power 'The first task of the new Severn Barrage Committee (to be set up on the suggestion of the Select Committee on Science and Technology) will be to review the further work needed and to make substantive recommendations to the Secretary of State for Energy. The Government has provisionally allocated £1.5 million for further studies in the light of these recommendations and intends that work should proceed as quickly as possible. Work on gathering the environment data needed for further feasibility assessments has already started, with the use of wave rider buoys to measure the height and frequency of waves in the Severn Estuary, at a cost of £42,000. The Government will keep in mind the possibility of worthwhile projects on smaller estuaries. It may also be possible, in some coastal locations, to recover the energy from tidal streams, as distinct from the hydrostatic heads developed in tidal basins. The Government will keep under review the potential contribution from other aspects of tidal power than the Severn Barrage alone.'

Geothermal Energy 'The Select Committee considered that the potential for geothermal energy to make a significant contribution to energy supplies was, on present information, very limited. The Government accepts that any contribution to energy supply from this source is likely to be long term, but, if its research programme can be carried forward to a successful conclusion, there is, nevertheless, the prospect of eventually securing worthwhile quantities of energy from this source.' (HMG White Paper 1978)

The DEn-supported *Energy Commission,* which had been set up by Tony Benn when Energy secretary as an open forum for discussion, also produced a report in 1977: see Table 2.1A. This foresaw a significant *possible* total contribution even in the short term, that is, 30–40 mtce by 2000 AD, even if it also argued that the actual contribution by that time would probably be much less (Energy Commission 1977).

In 1978, the Labour Government published its Green Paper on energy policy, which, although it continued to back a major (40 GW) nuclear programme (along with coal and conservation), saw renewable energy

Table 2.1 Early estimates of the potential of renewables by 2000

A. Data from energy commission paper 1

Energy source	mtce per annum
Solar space heating	1
Solar water heating	4
Biomass fuel	3
Waves	Up to 15
Tides	Up to 4
Wind	Up to 8
Geothermal	Up to 4
Total	39

The Energy Commission was an open Department of Energy-backed consultative forum for discussions relating to UK Energy Policy, introduced as part of the Labour Government's policy of open government (Energy Commission 1977).

B. Possible energy supply from alternative sources by 2000 AD (official estimates)

Energy source	mtce per annum
Solar: space heating	3
Water heating	3
Biomass fuel	3
Waves	Up to 15
Tides	Up to 10
Wind	Up to 8
Geothermal	Up to 4
Total	46

In million tonnes of coal equivalent (mtce)
Source: HMG Green Paper (1978)

sources as meriting careful review (see Table 2.1B). By this time ETSU had produced a full range of studies assessing each source, such as: Energy Paper 16 *Solar energy: its potential contribution within the UK* (1976), Energy Paper 21 *The Prospects for the Generation of Electricity from Wind in the UK* (1977) and Energy Paper 23 *Tidal Barrages in the Severn Estuary* (1977). Energy Paper 42, on Wave energy, was to follow in 1979.

Reports like these, and the public inquiry on the proposed expansion of nuclear fuel reprocessing facilities at Windscale in 1977 (which involved submissions on renewables by some opponents), led to increased coverage of 'alternative energy' in both the popular and technical press and to growing public and professional interest.

Support also came from the electricity industry. Glyn England, then the chairman of the nationalised utility, the Central Electricity Generating Board (CEGB), pointed out that in theory windmills could 'produce the equivalent of roughly a quarter of the CEGB's present output', while wave-power systems could 'supply the whole of Britain with electricity at the present rate of consumption' (England 1978).

In addition, there was growing pressure from environmental groups like Friends of the Earth for increased attention to be given to renewables (and energy conservation) as an alternative to nuclear power. This concern was also shared by a growing body of independent consultants and academic researchers, as reflected by influential technical studies which were produced at the time, as noted earlier (UK-ISES 1976; CAT 1977).

On the policy side, in 1979, the Centre for Alternative Industrial and Technological Systems at the North East London Polytechnic published my own report 'Energy Options and Employment' which tentatively claimed that a serious commitment to renewables and conservation could create perhaps twice as many direct and indirect jobs as an equivalent nuclear programme, and jobs in those areas and skills were hardest hit by the recession of that time (Elliott 1979). This argument was taken up by Labour movement organisations like SERA, the Socialist Environment and Resources Association. Some of the ideas that had emerged for alternative energy technologies were also included in the 'Alternative Corporate plans' developed in the mid-1970s by grass-roots trade union groups in the Aerospace and power engineering sectors (Wainwight and Elliott 1982).

The level of political enthusiasm at this stage, at least in some circles, might be judged by the suggestion by Frank Hooley, a backbench Labour MP with special interest in energy policy, in a response to the 1978 Green Paper, that a renewables' contribution to British energy needs of 100 mtce per annum by the year 2000 ought to be the aim. That said, the official view within the DEn and the Labour Government remained that nuclear power was the main new option for the future and that renewables were unlikely to make a significant impact in the foreseeable future, although they were still worth exploring as longer-term options. It fell to ETSU to try to put the situation in perspective.

In 1979, ETSU brought its various separate studies of renewables together into a more integrated assessment in Energy Paper 39 *Energy Technologies for the UK*, which set assessments of the various energy systems (both renewables and conventional technology) against an overall background of energy supply and demand forecasts.

It was quite severe. It concluded that 'in general it seems unlikely that, as a class, the renewable sources will be sufficiently competitive in cost on a large enough scale to displace coal or nuclear power from their central roles in the UK's energy supply'. But it added that 'some of the renewable sources promise to be cost-effective in parts of some markets, and their development is therefore justified as a way of offering a useful supplement to more conventional supplies, if for no other reason. More important, however, is the role of some of the renewable sources as an insurance if, for some reason, coal or nuclear power could not meet the demand for energy' (ETSU 1979).

The idea of 'insurance technologies' was developed as follows:

> The concept of insurance is important. Even if the country plans to achieve a particular pattern of energy supply and use, it might for a number of reasons find itself having to adopt some different pattern. Current assessment of future likely events and technical progress could be profoundly mistaken. There might be unexpectedly high world energy prices or restrictions on trade, including sudden embargoes, or restrictions on the use of certain energy sources. For example, nuclear power might be limited for environmental or social reasons, or concern about the effects of the increasing levels of carbon dioxide in the world's atmosphere could lead to limits being set on fossil fuel combustion well within a 50-year time horizon.

It went on 'Different technologies can provide a hedge against different risks. A robust R D & D strategy must ensure that technologies that might need to be used as an insurance are brought to an appropriate stage of development, so that if required they can play their insurance role. To be important as an insurance a technology does not have to be cost competitive with those with which it might have to compete in normal circumstances. Wave energy, for example, may never be competitive against nuclear power; but if for any reason nuclear power were not able to be used, wave energy could be an important alternative.'

Wave power was thus seen as a possible insurance technology, as were offshore wind and bioenergy and some of the others, since they could ultimately make fairly substantial contributions (e.g. up to 40 mtce per annum from wave power), if needed to replace some other options. However, it was considered more likely that the renewables would contribute simply on a supplementary basis. They would supplement the conventional sources and operate wherever cost-effective, essentially in a 'fuel saving' role, whenever the intermittent natural energy sources were available.

From a contemporary perspective, this approach clearly marginalised renewables as 'supplemental', with some being relegated to 'insurance' options, to be left until they might be needed in future, if other options were not available. Although rather conservative, given the situation at the time, with much uncertainty and tight budgets, this view might nevertheless be seen as a reasonably flexible and open-ended compromise.

EP39 was careful to avoid producing any estimates for the *total* potential for renewables, presumably because it would make little sense to simply 'add up' the various possible contributions from each source if fully developed (some would not or could not be). But they did provide individual estimates for the ultimately available theoretical potential for each individual source, as shown in Table 2.2. Note that Tables 2.1A and B were based on 'technically feasible' contributions by 2000 AD; Table 2.2 looks beyond that to the 'ultimate' possible contributors, what EP39 calls the 'potential ultimate market' for each source. Note the very large estimates for solar heat, seen as a major long-term resource.

Table 2.2 EP 39: Ultimate potential of UK renewables

Solar space heating	40–30 mtce
Solar water heating	10–20 mtce
Vegetable fuel	60 (but only 15 mtce as useful fuel)
Waves	10–40 mtce
Tides	4–10 mtce (Severn)
	2–3 mtce (others)
Wind land based	up to 10mtce
Coastal	up to 5 mtce
Offshore	0–20mtce
Geothermal	Several tens of mtce

Million tonnes of coal equivalent (EP39 uses the phrase 'potential ultimate market')

Source: EP39 (ETSU 1979)

The estimates are all essentially based on assessments of the scale of the physical resource available and the likely success and efficiency of practical technologies for harvesting the resource. The actual eventual contribution would depend on the rate of development and deployment of these systems, which in turn depended on how much was invested in them as opposed to other energy options, on the relative economics of the new systems when developed compared with conventional sources and on the renewables' success in challenging the existing fuels in the relevant market sectors. In addition, there would also be environmental and social constraints.

So, since all these factors interact, the actual contribution (or 'market share') that might be achieved in practice, depending in part on the policies and priorities adopted, could obviously be significantly lower than the 'potentials' given in Table 2.2. For example, EP39 concluded that on the basis of ETSU's standard future scenario (in which nuclear power continued to develop as the main new energy option), renewables would only contribute an annual total of 2.1 mtce by 2000 and 8.8 mtce by 2025. A meagre amount.

However, that was based on current policies and plans: obviously other priorities could emerge. EP39 included a preliminary report of an assessment being carried out by a joint ETSU/University of Aberdeen team of some of the alternative 'low energy' scenarios that had been produced by external bodies, based on a major renewable contribution meeting significantly lowered demand without reliance on nuclear power. EP39 commented that it seemed possible 'that several distinct low energy futures will be shown to be internally consistent and will be judged to be technically quite plausible in terms of their energy implications. The main areas of doubt seem likely to be the questions of economic and social consistency and plausibility' (ETSU 1979).

2.3 Economics to the Fore: ACORD Steps In

ETSU's EP39 report emerged just at the point when politics in the UK changed radically, with the election, in mid-1979, of a Conservative Government led by Margaret Thatcher dedicated to tight control of public sector spending and hostile to state intervention. As we shall see, the impact on the renewable energy programme was that economic concerns began to dominate over energy resource concerns: what mattered was not the scale of the resource but its cost.

By the end of 1982, some £32 million had been allocated to renewables R&D, and the next stage seemed likely to require even more funding. A decision on future priorities was evidently required, with likely unit costs being seen as the key issue. However, unit cost estimates for conceptual rather than real systems are obviously fraught with difficulties. Those for wave power seesawed dramatically. At one point there was talk of a cost of 50 pence per unit (kWh), clearly putting wave power out of court, a view that apparently emerged at major International Wave Energy Conference held at Heathrow in 1978 (Wilson 2010). But soon after, some estimates had dropped to below 10 p per unit (the initial target suggested by ETSU) and continued to fall, to within reach of a new target of 5 p/kWh. By 1981, some researchers estimated that 4 p/kWh or an even lower figure might be reached (see Table 2.3).

Wind-power estimates were consistently lower, and more solid, as a consequence of the quite large programme that had been underway in America since 1974. By 1981, it looked as if megawatt turbines might produce electricity at around 1.5 p/kWh ex-plant, given mass production, making them clearly economically attractive. Even with turbines mounted offshore, unit costs were being cited at around 2–3 p/kWh, competitive with conventional sources (Musgrove 1981), while the estimated unit cost of the Severn Tidal Barrage was not far off competitive, at around 3.8 p/kWh, according to Energy Paper 46 (Bondi 1981).

Table 2.3 Estimated unit costs for wave-power systems

	Cost in pence/kWh (Research team estimates)		
System	1978	1979	1980–1981
Salter's duck	20–40	10	4.5–5.6
Cockerell's raft	20–40	12	9
Oscillating water column	20–40	13	6–11
Lancaster airbag	5–10	6	3.5–4.7
Bristol cylinder	Not quoted	14	5–10
Belfast buoy	Not quote	12	8
Lanchester clam	Not quoted	6	4–4.5

Ross (1981)

In 1981, the DEn produced a report for a UN conference on renewables, which summarised the 'state of play' in the UK at that time. In terms of the contribution that might be expected in the UK by 'early next century', it put the geothermal resource at 5 mtce per annum, solar 5, solid biomass 6, liquid biomass 6, wind 5, tidal 10, hydro 3 mtce (there already being 2 mtce), but wave only 'small'. Tidal was still seen as a front runner. The report noted that 'a pre-feasibility study of the Severn Estuary suggests that a barrage might produce electricity at a cost not far above that of a coal-fired station'. By contrast, on wave power, it noted that 'the central problem is whether costs can be brought below 5 pence/kWh' (Department of Energy 1981).

Although John Moore, then an Energy Minister, had claimed at the opening of a large wave energy test tank at Southampton in Sept 1980, that 'whatever other problems our wave energy researchers may face, lack of Government support will not be among them' (Moore 1980), evidently doubts were beginning to emerge about wave power. Indeed, there were reports that the DEn was 'planning to withdraw from funding wave power projects' (Fishlock 1981; Ezard 1981), following a comment by the then Secretary of State (David Howell) that the renewables R&D programme was to be 'critically reviewed' the following year.

Responding to these reports, the then DEn Chief Scientist, Dr Anthony Challis, wrote, in a letter to the *Financial Times,* saying that the wave-power programme had 'produced a large number of possible solutions and there are grounds for continuous optimism about the feasibility of utilising this resource. The next phase, however, would involve work on a larger scale and this would be very costly. The Department would therefore wish to concentrate its efforts on a smaller number of devices.' Hence, there was a need for an assessment exercise to 'decide whether or not to proceed with any of the devices which form part of the present programme' (Challis 1981).

In 1980, the Department had in fact cut back on the number of wave-power projects it was funding, from nine to four, with further funding for two of the early entries, Salter's nodding duck and Cockerell's raft, also being halted. Now the remaining systems were to be re-assessed. According to David Fishlock, writing in the Financial Times, 'the four teams still being funded by the Department of Energy have until March 1982 to show that wave-power machines may achieve electricity costs as low as 5 p a kilowatt hour at present prices'. But he added 'The Department expects that none of the four systems in test will succeed in this' (Fishlock 1981). However, in his FT letter response, Challis was insistent that no decisions had been taken and that 'I and the experts who advise the Secretary of State for Energy must keep an open mind until the results of the present phase of work are available.'

The research teams themselves were unhappy that they were being forced to produce cost estimates on such a short schedule and at such an early stage of development—the wave programme has only formally started in 1976. Stephen Salter, the inventor of the 'nodding duck', felt that it was like trying 'to decide our aviation policy on the data available in 1910' (Salter 1981). Nevertheless, the review went ahead, and, as we shall see, it resulted in some dramatic changes.

ACORD, the Governments Advisory Council on energy R&D, had been charged with making the assessment, as part of its regular round of reviews of energy technology R&D, with ETSU's help. Some of the results emerged in March 1982. It transpired that wave power had been all but abandoned, along with active solar, biogas and offshore wind. There was considerable clamour for clarification and in particular calls for the publication of ETSU's background documentation. Eventually these were acceded to.

The way in which this affair was managed drew as many criticisms as the actual conclusions. As we have seen, the demise of wave power, despite the £15 million that had been spent by that stage, had been rumoured some time before the ACORD review. But even after the initial news leaked out, following the ACORD meeting on 19 March 1982, that it has indeed been unfavourably assessed, there were reports that a final decision on wave power would be delayed until September, in order to take account of new data.

As *Engineering Today* put it: 'An unexpected revolt by some ACORD members against the civil service advice led to a six-month reprieve for the programme. But the decision on how to deal with the recommendation has still to be made by politicians and their Civil Service advisors' (Engineering Today 1982a). Electrical Review reported that 'The Council decided to urge the Department of Energy to give the four key teams another six months so that they would be able to examine and answer a consultant's report which prices all of them out of the 5 p/kWh limit set by the Department. Some of ACORD members were angry to discover that the teams had not even been shown the report drawn up by the Government's consulting engineers, Rendel, Palmer and Tritton' (Electrical Review 1982a).

The confusion was at least partly resolved when the DEn issued a 'Summary of Advice' on 2 April 1982 (Box 2.4), even if there was some speculation that the summary did not reflect all that ACORD had recommended in their full unpublished report. There were reports suggesting that ACORD members had not been consulted on the wording of the summary and that some at least were unhappy about it. For example, the idea of a 'reprieve' for wave power had evidently more or less evaporated. Some disquiet was also expressed concerning the fact that the ETSU project managers had not been invited to attend the ACORD meeting (Ross 1985, 1995).

These issues were subsequently raised in a parliamentary question on 28 June, which 'asked the Secretary of State for Energy if he was satisfied that the summary of advice proposed by the Advisory Council on Research and Development adequately reflected the views of those serving on the council; and whether the views of those involved in renewable energy projects were excluded from the council's consideration'.

David Mellor, then an under Secretary of State for Energy, responded: 'Yes. There was no question of excluding the views of those involved in my Department's renewable energy projects. The reports of all the renewable energy steering committees were made available in full to the council and the chairman of the committees attended and took part in the discussion' (Mellor 1982a). So the chairman of the steering committees had attended, but not the project managers.

As can be seen from the summary in Box 2.4, ACORD recommended that no further work should be funded in the wave power, active solar, biogas systems or offshore wind fields: existing results should simply be written up. But no details of future funding were cited. The impression was that it was now up to the government, via the DEn, to consider ACORD's advice.

Box 2.4 ACORD Advice Summary (Department of Energy 1982)

Advisory Council on Research and Development for fuel and power (ACORD)

'Summary of advice to the Secretary of State for Energy on his research and development programme on renewable energy sources

1. In conducting its review, the Council kept to the fore the three principles underlying the Department's Renewable Energy R & D policy:

   (i) that during the early stages of the programmes when the techno-economic uncertainties surrounding technologies were high – as many resources technologies would be investigated as could sensibly and appropriately be accommodated within the available funds and manpower resources;
   (ii) that as the uncertainties surrounding the prospects for economic development of technologies reduced through R & D, those which showed poor prospects of becoming economically competitive against other technologies expected to be available, would be dropped from the programme;
   (iii) that only research and early development work not normally beyond the first prototype stage – would be considered as potentially appropriate calls on the Department's R & D budget. Programmes were recommended on the basis that they would be supported at minimum cost to the public funds commensurate with considerations of efficiency, effectiveness and value for money.

2. General Recommendations. At its review this year the Council assessed the programmes against the above principles and made three general recommendations. These recommendations were:

   (i) the time had come to abandon R & D into the less promising technologies and to continue with the more promising ones still requiring R & D. Available financial and manpower resources should be directed towards the development of these technologies on a selective basis.
   (ii) The budget ceiling, approximately £11 m per annum at out-turn prices for the next 2 to 3 financial years on the basis of current estimates, would be sufficient to support the R & D work the Council recommended should be continued and at the pace the Department considered was necessary at this time.
   (iii) As technologies developed beyond the R & D stage, the burden of financial support and responsibility for later stages leading to commercialisation should be taken up by other Government Departments or public authorities having the appropriate responsibility, and the private sector.

3. The Council took the view that sufficient information was now available to allow some categorisation of the technologies and made the following recommendations on each of the Department's renewable energy programmes.
4. Specific Recommendations

   (1) *Passive solar design:* A modest amount of money should be devoted to selected further work on research into passive solar design. However, the thrust of the Department's available effort should be directed towards encouraging greater involvement by the Department of the Environment, the building industry and related professional and financial institutions to ensure that promising lines are followed up via demonstrations and other work leading to application.
   (2) *Solar water heating:* No new work should be started on solar water heating for the UK. It would be desirable for existing contracts to run their course. If, at a later stage, developments merit it, the Department should consider the scope for supporting further work in this area.
   (3) *Active solar space heating:* No new work should be started on active solar space heating. Existing work should be completed, to allow the results to be published. The Council took the view that the Department ought to try to achieve an orderly rundown of solar water and active space heating for an expenditure of less than £0.5 million.
   (4) *Geothermal aquifers.* Budget permitting, the Department's programme should include the drilling of a further production size borehole in an unexplored basin with the aim of providing heat for an identified market should the borehole locate a suitable aquifer, in addition to providing data. If the budget did not permit it or if a market could not be firmly identified, the drilling programme should be deferred. The Council supported the Department's efforts to ensure that future exploitation of aquifers by local authorities for other than R, D & D purposes should be the responsibility of the Department of the Environment.
   (5) *Geothermal hot dry rock.* Work should continue broadly along the lines planned in the proposed programme. However, the next phase would be substantially more expensive. Consideration would therefore have to be given at the appropriate time to whether the Department could afford to investigate the prospects for hot dry rock technology at greater depths.
   (6) *Biofuels: combustion:* Biofuels combustion R & D should continue broadly along the lines planned at or about the proposed level of expenditure.

(7) *Biofuels: anaerobic digestion of animal and vegetable materials.* The work should continue broadly along the lines planned at or about the proposed level of expenditure.
(8) *Biofuels: thermal processing.* No further work on biomass (energy crops) gasification leading to methanol should be supported by the Department. It would be advisable for existing work to be completed to allow the results of this part of the programme to be published if it is decided to do so.
(9) *Onshore wind power.* The horizontal axis and vertical axis parts of the programme should continue as planned. Attempts should be made to secure a financial contribution from the Department of Industry for the vertical axis studies. Existing national resources size studies should be completed and no new work started.
(10) *Offshore wind:* It would be desirable for the offshore studies to continue at the minimum level required to keep options open for worthwhile international collaboration and that opportunities for such collaboration, as a means to minimise the Department's financial burden, should be explored. If necessary, however, this work could be dropped from the programme for the time being.
(11) Wave *power:* No new development work on wave power should be supported from the Department's R & D budget: however, a detailed comparison of the major systems was planned for September *1982* when most of the present work is scheduled for completion. The existing contracts should be allowed to run their courses. The results from all the projects should be prepared for publication by the Department. It was important, in the Council's view, to present the results in a tidy form so that the work supported to date would be available to Government and private sector should it be decided to re-consider the role of wave power in the UK energy economy. If necessary, limited funds should be made available to support this activity. The Council recommended that if the budget did not permit both the existing contracts to run their course and the results of the work to be prepared for publication, priority should be given to the latter.'

It should be noted that tidal barrages were not included in the ACORD review, since they were the subject of a separate review chaired by Sir Herman Bondi (Bondi 1981). Some of the results of that, and subsequent barrage reviews, will be discussed in the next chapter. They were mostly favourable, but, at this stage, barrage technology, which was similar to that used for hydro, was not seen as needing significant R&D funding, unlike most of the technologies looked at in the ACORD review. It should also be noted that the potential for expansion of hydro was generally seen as limited in the UK, although smaller projects could make a useful contribution.

2.4 Reactions to the ACORD Review

The renewable energy research community was horrified by the outcome of the ACORD review, particularly in respect to wave energy, but it was assumed that there would be a period for consultation before a final decision was taken. Norman Bellamy of SEA-Lanchester (involved with the 'Clam' wave energy project) said: 'Let us hope that Mr Nigel Lawson [the then Secretary of State for Energy] rejects what the ACORD report says about wave power and has the foresight to seize the opportunity which wave energy offers. If the Government does not back the development, there is a real danger that this British project will be taken over and developed by other countries – with the support of their Governments' (Bellamy 1982).

However, it was not to be. Indeed, it become clear that the total budget had in fact been established long before the ACORD review, and that it was part of ACORD's brief to make cuts to fit it. That clarification emerged accidentally at a conference organised by the Institution of Mechanical Engineers on 21 May in London. This was, as one speaker noted, the first major gathering of renewable energy researchers since the ACORD recommendations had been published. According to Professor Brinkworth, of Cardiff University Solar Centre, the recommendations were of a 'precipitative nature'. He evidently hoped the Secretary of State would adopt a more flexible and enlightened approach, and not accept what he saw as ACORD's 'bad advice' (Electrical Review 1982b).

At this point, according to a report at the time, an understandably somewhat tremulous DEn spokesman stood up and told the audience that in fact the decision had already been taken (NATTA 1982). The £11–12 million ceiling had been set long before ACORD's review and the Department would shortly be announcing specific programme budgets. It transpired that the decision to reduce overall spending to £11–12 million, since 'it will not be possible to insulate this area from the savings the Government are making in public expenditure', had been first disclosed, but not widely appreciated, in a parliamentary answer on 31 March, and repeated on 17 May 1982, when Energy Minister David Mellor added that the ACORD recommendations were 'acceptable to the Department' (Mellor 1982b).

Subsequently in June 1982, Mellor outlined the specific 1982–1983 budgets—£1 million for solar, £3 million for 'deep' geothermal hot dry rocks (HDR), £3 million for aquifers, £3.2 million for wind and £3 million for 'completing' the wave-power work (concluding evaluations and writing up results). He added 'no new development work will be initiated' on wave power, and that the active solar and biomass programmes were to be cut.

At a meeting later in June 1982 of the all-party Parliamentary Liaison Group for Alternative Energy Strategies (a forum for MPs and the renewable energy research community), Mr Mellor, facing a somewhat hostile audience, outlined the rationale for the cuts as follows: 'The decision I announced in the House on 17 May in no way constitutes a termination of work on renewables [...] Rather, this is a decision to move on in our programme by concentrating our efforts on the most promising technologies. This has been inherent in all our work since the programme began seven years ago – sooner or later the field would have to be narrowed down' (Mellor 1982c).

However, Mellor did not let ACORD off the hook too easily: 'Our economic policy requires that we impose constraints on public expenditure. But the fact that these constraints led us to conclude that departmental funds for renewables R & D would probably be limited to £11–12 million a year for the next few years did not restrict ACORD in making its recommendations. The council was at liberty to recommend that this sum was clearly inadequate for the maintenance of a proper programme. It decided however that these funds were sufficient to conduct the proposed work on more promising technologies at the pace considered appropriate at this time.'

Faced with what might be seen as a fait accompli, there was little the renewable energy research community could do but complain. The bulk of their criticism subsequently moved from the 'public handling' of the news about the decision and on to the internal assessment procedures ACORD had used, with its treatment of wave energy being a key focus. It was widely remarked that it was strange that the ETSU programme managers had, evidently for the first time, been excluded from the ACORD meeting. Clive Grove Palmer, the ETSU wave-power programme manager, subsequently took 'early retirement', at least in part because he was, he said, 'fed up' with the way in which the programme had been handled, although he insisted that he was 'not walking out in a huff' (Engineering Today 1982b).

For their part, the wave-power researchers were particularly incensed that what they alleged to be inaccurate and out of date unit cost estimates had been used in the assessment. Gordon Senior, a consultant involved in providing ETSU/the DEn with analysis, complained that his conclusions had been modified (or even reversed). Recriminations and assertions about alleged errors continued to emerge, with Senior and Stephen Salter submitting critical comments to a subsequent House of Lords Alternative Energy review, while ETSU and the DEn submitted rebuttals (House of Lords 1988).

The latter, interestingly, included a rebuttal of suggestions that the location of ETSU within AEA Harwell had any significance. ETSU was outside of Harwell's security perimeter and, although some of its staff did switch between nuclear and renewables work, some had been recruited from outside and were not involved with nuclear work.

However, complaints, denials and rebuttals apart, back in 1982, with only the ACORD summary formally available at that stage, the focus of the reaction to the review then moved to the confidential ETSU background documentation. Initially, the Secretary of State for Energy had refused to make these internal reports public. But after repeated calls, ETSU's report 'Strategic Review of renewable energy technologies' (R13) was eventually released in November 1982, followed later by 'Contribution of renewable energy technologies to future energy requirements' (R14).

The ETSU reports R13 and R14 proved to be a mine of useful information. ETSU had done a considerable amount of interesting work. For example, R13 developed on EP39's long-range estimates of future potential contributions from each source based on the estimated scale of the resource and also provided unit cost estimates. The results are summarised in Table 2.4, which is abstracted from R13.

Of course, as noted earlier, it is a long way from potential contributions to technically, environmentally and economically viable systems.

Table 2.4 Cost and supply data from ETSU R13

Electricity production/ultimate potentials as a percentage of current annual electricity consumption, in terawatt-hours per annum, plus cost estimates in pence/kilowatt hour		
Offshore wind	50% (149 TWh per annum)	3.1–7 p/kWh
Onshore wind	20% (50 TWh per annum)	1.9–4.3 p/kWh
Tidal (all sites)	15% (13 TWh per annum)	2.8–3.1 p/kWh
Wave power	25% (66 TWh per annum)	4–12 p/kWh
Geothermal hot rocks	10% (2500TWh over 100 years)	4–5.2 p/kWh
Small wind	10% (25 TWh per annum)	–
Micro-hydro	1% (1.8 TWh per annum)	–
Heat production/ultimate potential in million tonnes of coal equivalent (mtce) per annum:		
Solar		
	Passive	3
	Active	4
Biofuels		
	Vegetable	3–4
	Animal	3
	Thermal processing	5 (methanol)
	Combustion of refuse	12–14
	Others	5
Geothermal	aquifer	0.5–l.5
The ultimate potentials cited here are the technically feasible potential contributions leaving aside economic and environmental constraints		

ETSU (1982a)

The overall conclusion was that nevertheless: 'on current cost estimates all of the renewable sources examined show some prospects of being cost effective during the review period [by 2025], but that [...] all the technologies seem to need some sort of Government backing to help them get established'. Without that, progress would obviously be less spectacular.

So given that renewable research would have to compete for funds with other options, what would be the actual contribution of renewables in reality? That would obviously depend on the overall level of demand for energy, the economics of the various options, environmental constraints and of course on government energy policy.

ETSU's R14 'Contribution of renewable energy technologies to future energy requirements' explored the likely rate of diffusion of the various renewable energy technologies, using the ETSU supply/demand/ fuel price scenarios (up to 2025 AD) as a backdrop. The conclusion was that 'the technical potential of the renewable resources is about half to two thirds of the total UK energy supplied by oil, coal, gas and nuclear. But these are unrealistic figures since they take no account of the economic, environmental and market related factors' (ETSU 1982b).

Thus by 2025, wind, tidal and geothermal power were seen in reality as, together, supplying at most, on the best scenario for renewables (i.e. assuming high fuel prices) only 5.8% of Britain's electricity, with a maximum on-land wind energy installed capacity of 7.5 GW and only 4 GW offshore. It is interesting to compare these projections with the much higher estimates produced by the CEGB. For example, in 1981, it put the 'ultimate' offshore wind potential at 70 GW. Although in its 1982 evidence to the Sizewell public inquiry, it concluded that, in the short term, 'none of the renewable technologies could at present offer with any assurance an economic, reliable and sufficient source of energy to be competitive with nuclear power', it claimed that renewables might prove attractive in the longer term, possibly contributing about 15% of Britain's grid electricity by 2030 'without major system operational difficulties', rising to 20% if geothermal power was included (Wright 1982).

For comparisons sake, the *current* (2018) renewables' contribution is actually around 30%, with 40 GW installed, including 7.6 GW of offshore wind. With hindsight, it seems the CEGB was nearer the mark than ETSU, though still offering projections for 2030 which were well below what has turned out to be possible long before that. It is also perhaps worth noting a comment made by John Wright, in a 1981 CEGB review booklet on renewables, that solar power 'is never likely to be of much use for supplying electricity to the Grid in this county'. It is now at 12 GW.

However, in the early 1980s, the establishment view of what might be possible was relatively cautious. ETSU's revised conclusions and estimates were reflected in the new classification system for the technologies that was presented in R13 (see Table 2.5). These classifications reflected their final assessment based on all the factors and constraints, but with cost-effectiveness well to the fore.

As can be seen from Table 2.5, on-land wind and tidal power emerge as front runners in Category A (which R13 labelled as 'strongly placed'), wave power being relegated to Category D ('long shots'), along with

Table 2.5 Summary categorisations (A–D) of renewable energy technologies from ETSU R13

	A	B	C	D
Heat producers		Passive solar space heating Geothermal aquifers	Solar water heaters Geothermal hot dry rocks (2)	Active solar space heating
Electricity producers	Onshore wind: Tidal	Small-scale hydro (1)	Offshore wind: Small wind (3)	Wave
Fuel producers		Biofuels: Combustion Anaerobic digestion of animal wastes	Biofuels: Anaerobic digestion of vegetable wastes Thermal processing	

ETSU (1982a)

Category A includes technologies that are economically attractive now or in the near future and for which a route to their exploitation, without too serious difficulty, can be envisaged

Category B technologies are also economically attractive, but there may be factors mitigating against market uptake

Category C technologies are generally further away from the deployment stage. They might be less cost-effective than competing renewable energy technologies or they might only be cost-effective at a future date. Improvements in the cost or performance resulting from further R&D may improve the prospects for such technologies

Category D covers technologies which the analysis shows may be cost-effective only under the most favourable circumstances (i.e. with very high costs of conventional energy) and in the very long term. In some cases (1) little further R&D is required, in others much further information is required on the technology (2) or on the market features (3) before an adequate assessment of the potential is possible

active solar power, although, as far as wave power was concerned, R13 'strongly recommended that no decisions are made before the detailed cost review has been completed'.

Even so, wave power was seen as likely to be expensive. R13 puts it as follows: 'Wave power is likely to be economic only in those futures more favourable to renewable energy technologies […] Although wave power could just be economically acceptable at the bottom of its estimated cost range … other electricity generating sources… are consistently more attractive when analysed under the same circumstances.'

Of these 'the Severn Barrage generally has the best economic prospects of all the renewable sources, bettered only by on-shore wind power on the latter's lower cost. In general, tidal power is roughly on a par with nuclear power in the benefit/cost ratios which it produces.'

The main criterion was economic cost, but R13 did attempt to explore non-economic factors, by which it meant the organisational, legal and environmental problems that might hinder market take-up of the technology. Even so, it was the market mechanism which would be the final arbiter. However, R13 did hint that there might be 'imperfections' in the market mechanism, market inertia, for example, or the non-availability of a fully developed industrial manufacturing infrastructure ready to take on the new technologies: 'in most cases effective supply structures do not yet exist'.

The DEn programme was therefore designed in part 'to act as a temporary focus for the technologies in the period until commercial supply structures begin to emerge. Slowly, these structures are indeed beginning to crystallise out, though only with Biofuels from wastes to a discernible degree, and even there only in very preliminary forms.'

Whereas technologies like passive solar power might interest and feed into a diverse market ('house purchasers, builders, architects local authorities town planners'), it said that onshore wind and tidal power 'have the great advantage of being able to look towards large and well organised customers in the form of the Generating Boards'. In other words, they could be left to the CEGB.

Addressing the problem that new sources faced in getting established, in R13, ETSU pointed to the need for 'demonstration projects': 'To help overcome what could be categorised as market "inertia", it might be appropriate for Government to support demonstration projects. These would involve full-scale renewable energy devices being installed to operate in exactly those conditions which the technology would encounter in its lead markets, with the costs being borne partly by a selected end user and partly by Government. If successful innovation is to be fully exploited and the benefits of R & D reaped, then the necessary effort must be put into the following through from development into demonstration. "Demonstration" – showing the benefits of the technology to potential users – has therefore come to been seen as an integral part of the R & D process. In an area as central to the well being of the UK as energy supply, it could be that Government would wish to play a significant part in supporting appropriate renewable energy technology demonstration projects and their subsequent promotion in this way.'

However, ETSU drew back from calling for more direct market stimulations via subsidies. It said that grants and subsidies to stimulate the market for renewables do 'not figure prominently in the analysis... because proper consideration of such powerful weapons calls for analysis of their macroeconomic effects which is beyond the scope of the review. In fact the idea of market adjustment has only been introduced for one technology [solar water heating] and then more as food for thought than as a conclusive recommendation.'

ETSU was perhaps reflecting the then dominant view (with a conservative administration in power) that the Government's responsibility extended only to research and development, and perhaps demonstration. Once proved to be cost-effective, and therefore potentially commercial, the technologies should be able to stand on their own feet, that is, be taken up as economic propositions by private (or public) industry.

ETSU was evidently not totally convinced by this argument. For example, it hinted that the technologies that might get developed in this way may not be those most important in national terms: 'It should be stressed that it was the objective of the market analysis to make a realistic assessment of each renewable technology's potential to penetrate its markets, giving full weight to the fact that market response is determined primarily by the investment criteria of end users in the market place. This market share might, of course, differ from the nationally desirable market share determined by economic analysis based on national investment criteria. The comparison between the national and market views of prospects was central to deducing the further research, development demonstration or deployment activities needed to help achieve market penetration, and to defining Government's role in these processes' (ETSU 1982a).

However, as we have seen, it drew back from proposing any significant government intervention, which is hardly surprising given the economic philosophy of the Conservative Government. Nevertheless, the need to consider government funding beyond the RD&D phase was hinted at by Dr Freddy Clarke, who, from 1976, was AEA Harwell's research director, and in overall charge of ETSU. He had also been chair of ETSUs wave energy steering committee. In a book on Science Policy, he said 'It is perhaps a pity from the viewpoint of a technical innovation that government funding had to be restricted at the point of demonstration, when funding overall normally has to increase quite dramatically' (Clarke 1984).

As can be seen, R13 and R14 set the ACORD review in a wider context, and indeed, in hindsight, they were very perceptive about the strategic issues that would face renewables in the future. Even so, the end result was that wave power, and several other renewables, were deemed to have been assessed and failed. Nevertheless, for good or ill, the debate over the wave-power decision continued: indeed, as we shall see, it continued to resurface regularly. There were also calls for changes to ACORD. The 1984 Select Committee on Energy made suggestions for restructuring it and the associated review and assessment procedures, following criticisms that, for example, ACORD's membership was weighted too much in favour of the 'conventional' energy industry. Some restructuring was subsequently set in hand (Select Committee on Energy 1984).

The ACORD review had certainly set the, now reduced, programme a new direction, and arguably a more economically sensitive one. Wind power had come out of it relatively well, as had geothermal, while tidal barrages (as noted above, excluded from the ACORD review) were also still favoured, with £250,000 allocated for further desk studies on the Severn Barrage, matched by the same amount from industry. By contrast, in something of an obituary, the Wave energy programme results were duly written up in ETSU report R26, which concluded that 'there was only a low probability of any device achieving an energy cost below 8p/kWh' (ETSU 1985). That conclusion was still not shared by all, with, as has been covered in even more forensic detail by Wilson (Wilson 2010), the debate over wave power being long and tortuous.

However, as we shall see, the surviving technologies continued to make progress. Some significant projects emerged, as industrial companies become more involved, with ETSU providing guidance and programme management and ACORD carrying out further assessments. In effect, for good or ill, the initial exploration phase was seen as being over, and, for the survivors, the emphasis now moved on to commercial prototypes and industrial development.

References

Bellamy, N. (1982). SEA-Lanchester Wave Energy Brief No. 12, Lanchester Polytechnic, Coventry.

Bondi, H. (1981). *Tidal Power from the Severn Estuary* (Energy Paper 46). London: Department of Energy.

CAT. (1977). *An Alternative Energy Strategy for the UK*. Centre for Alternative Technology, Machynlleth, At That Point Know as the National Centre for Alternative Technology.

Challis, A. (1981, May 27). Work on Wave Energy. *Letter to the Financial Times*.

Chapman, P. (1977, October). Alternative Energy Sources; in 'Nuclear Power and the Energy Future'. *Proceedings of Royal Institution Forum*.

Clarke, F. (1984). Energy. In M. Goldsmith (Ed.), *UK Science Policy*. London: Longman.

Department of Energy. (1981). *Renewable Energy in the UK*. UK National Paper Prepared by the UK Department of Energy for the United Nations Conference on New and Renewable Sources of Energy.

Department of Energy. (1982, April 27). *Summary of Advice to the Secretary of State for Energy on His Research and Development Programme on Renewable Energy Sources*. Advisory Council on Research and Development for Fuel and Power (ACORD), Department of Energy.

Electrical Review. (1982a, April 2). Wave Programme Still Afloat. *Electrical Review*.

Electrical Review. (1982b, June 25). Renewable Energy Seeks Friends in Whitehall. *Electrical Review*.

Elliott, D. (1979). *Energy Options and Employment*. Centre for Alternative Industrial and Technological Systems, North East London Polytechnic.

Energy Commission. (1977). *Energy Commission Paper 1*. Department of Energy, London.

Engineering Today. (1982a, April 5). Wave Energy Scheme Rides Out the Storm but Is not Yet Afloat. *Engineering Today*.

Engineering Today. (1982b, May 3). ACORD not Given to Renewable Energy. *Engineering Today*.

England, G. (1978, October). Renewable Sources of Energy – The Prospects for Electricity. *Atom*, *264*, 270–272. UK Atomic Energy Authority.

ETSU. (1979). *Energy Technologies for the UK* (Energy Paper 39). Energy Technology Support Unit, Department of Energy, London.

ETSU. (1982a). *Strategic Review of Renewable Energy Technologies*. Energy Technology Support Unit, Department of Energy (R13).

ETSU. (1982b). *Contribution of Renewable Energy Technologies to Future Energy Requirements*. Energy Technology Support Unit, Department of Energy (R14).

ETSU. (1985). *Wave Energy*. Energy Technology Support Unit, Department of Energy (R26).

Ezard, J. (1981, May 14). Scientists Expect Wave Power Research to be Cut or Stopped. *The Guardian*.

Fishlock, D. (1981, May 13). Wave Power Studies May be Dropped. *Financial Times*.

Flood, M. (1986). Renewable Energy in the UK: Anatomy of a Government Programme, Case Study for an OU Course, T302 Design and Innovation, Block 4, The Open University, Milton Keynes.

HMG Green Paper. (1978, February). *Energy Policy*. Consultative Document, HM Government, Cmnd 7101.

HMG White Paper. (1978). White Paper on Alternative Energy Sources, Reply to the Third and Fourth Reports from the Select Committee on Science and Technology Sessions 1976–77, HM Government, London.

House of Lords. (1988). *Alternative Energy Sources*. House of Lords Select Committee on the European Communities, Session 1987–88, 16th Report.

Mellor, D. (1982a, June 28). Parliamentary Answer on ACORD, House of Commons, Hansard Col. 26 cc227-8W. https://api.parliament.uk/historic-hansard/written-answers/1982/jun/28/advisory-council-on-research-and#S6CV0026P0_19820628_CWA_122

Mellor, D. (1982b, May 17). Parliamentary Answer on Alternative Energy, House of Commons, Hansard, Vol. 24 cc17–8. https://api.parliament.uk/historic-hansard/commons/1982/may/17/alternative-energy#S6CV0024P0_19820517_HOC_152

Mellor, D. (1982c, June 29). PARLIGAES Meeting, Parliamentary Liaison Group for Alternative Energy Strategies, House of Commons.

Moore, J. (1980, September 26). Speech at the Opening of the Southampton Wave Test Tank, Department of Energy Press Release 191. Cited in Ross (1995) op cit.

Musgrove, P. (1981, April). Data Cited at a PARLIGAES Conference, Parliamentary Liaison Group for Alternative Energy Strategies, London.

NATTA. (1982). Report of the IMechE Meeting in a Supplement to NATTA Newsletter 17, May–June.

Ross, D. (1981). Wave Energy Newsletter, No. 3, May, London.

Ross, D. (1985). Wave Energy Newsletter, No. 5, May, London.

Ross, D. (1995). *Power from the Waves*. Oxford University Press.
Ryle, M. (1977). Economics of Alternative Energy Sources. *Nature, 267*, 111–117.
Salter, S. (1981, May 18). Work on Wave Energy. *Letter to the Financial Times*.
Select Committee on Energy. (1984). *Energy R, D&D in the UK*. House of Commons Select Committee on Energy, Session 1983–84, Ninth Report.
Select Committee on Science and Technology. (1977). *Alternative Energy Sources*. Select Committee on Science and Technology Sessions 1976–77, House of Commons Paper 5341/11/111, London.
UK-ISES. (1976). *Solar Energy: A UK Assessment*. UK Section of the International Solar Energy Society, London.
Wainwright, H., & Elliott, D. (1982). *The Lucas Plan: A New Trade Unionism in the Making*. London/Nottingham: Alison and Busby. New Edition, Spokesman Press, 2017.
Wilson, J. C. (2010). *A History of the UK Renewable Energy Programme, 1974–88: Some Social, Political, and Economic Aspects* (PhD Thesis). University of Glasgow. http://theses.gla.ac.uk/3121/1/2010WilsonJohnPhD.pdf
Wright, J. K. (1982, November). *Alternative Methods of Electricity Generation*. CEGB Proof of Evidence to the Sizewell B Public Inquiry, Central Electricity Generating Board.

3

The Technology Moves on *in the 1980s*

3.1 Industry Takes an Interest

From 1983 onwards, following the ACORD review, the emphasis shifted to the wind, tidal and geothermal programmes, with some hardware projects going ahead. The wind-power programme continued to develop with significant public and private capital being allocated. Particular interest was shown by private firms like Taylor Woodrow, GEC and McAlpines, working in industrial consortia like the Wind Energy Group (WEG), which was a joint venture backed by British Aerospace, GEC and Taylor Woodrow, or independently, like Glasgow-based Howden & Co. who broke into the medium-scale wind turbine market.

The WEG had built a 250 kW prototype on the Orkneys, and this was followed in 1987 by a flagship £10 million, 3 MW WEG wind turbine, with most of the funding coming from the DEn. This was much larger than anything that existed anywhere in the world at the time: the US Aerospace-led programme had been struggling with (and eventually abandoned) 2.5 MW machines. They were arguably too big too soon. However, the UK programme pushed ahead: there were plans for a further two 3 MW machines on the Orkneys. Smaller wind turbines were

D. Elliott, *Renewable Energy in the UK*, Energy, Climate and the Environment,
https://doi.org/10.1007/978-3-030-04765-8_3

also developed by WEG, with backing from the Department of Trade and Industry, including a 250 kW unit, sited in North Devon and designed for export. A 300 kW unit was also erected by James Howden & Co. on the Orkneys, alongside WEG's 250 kW prototype.

In 1985, Howden had made a significant breakthrough in the American market, installing 10 machines on a Californian wind farm. A 750 kW machine was also planned for the Shetlands. Work also continued on Musgrove's unique vertical axis machine, backed by McAlpines, with a 100 kW version being built alongside a 200 kW American-designed conventional horizontal axis machine erected for the CEGB by Howden on a power station site on Carmarthen Bay. The CEGB also planned to build a megawatt-sized machine at a site in Lincolnshire.

Following the generally positive Bondi review (Bondi 1981), interest in the Severn tidal barrage continued, with rival industrial groups proposing different schemes. One was the original 10-mile-long 8.6 GW barrage, as proposed in the Bondi report (to cost £6 billion and generate 6% of Britain's electricity) backed by GEC, McAlpine, Taylor Woodrow, NEI and Balfour Beatty. A similarly scaled rival scheme, also on the Bondi line, was proposed by the engineering consultants Rendel, Palmer and Tritton, backed by the Costains, Tarmac and HBM consortium (who built the Thames Barrier). They claimed it could be built for around £4 billion. In addition, a 1 GW mini-barrage, close to the (first) Severn Bridge, costing £885 million, and producing 1% of UK power, was backed by construction company Wimpey and the engineering consultancy Atkins. A 700 MW barrage on the Mersey was also proposed by an industrial consortium, the Mersey Barrage Company.

Geothermal Hot Dry Rock work, for power production, also prospered with the announcement in June 1983 of a further allocation, in addition to the £10 million already spent, of £11 million over three years for further work in Cornwall, led by the Camborne School of Mines. An aquifer system, for heat production, was also backed in Southampton. Solar Photovoltaics continued to be played down, although BP did construct a 33 kW test unit at the CEGB's Marchwood Engineering Laboratory near Southampton.

Wave-power research, however, was hard hit by the cut-backs. As we have noted, some £15 million had been allocated to wave-power research by 1982, with funding rising to a 'high' of £4.4 million in 1981–1982, 30% of total government expenditure on renewables, only to be cut back dramatically (down to £400,000 by 1984–1985) following the ACORD review. The cut-backs had a dramatic effect on the wave energy research community. Most device teams had to reduce manpower and scale down operations considerably. Some, like Wavepower Ltd. in Southampton, gave up entirely. But despite these funding setbacks, some wave-power research continued.

An optimistic interpretation might be that the remaining teams were now 'leaner and fitter'. The emphasis shifted from large gigawatt-scale deep-sea systems to smaller in-shore coastal and shore-line units for isolated island communities (Flood 1985). However, they were competing with Japan and Norway, who have made fairly rapid progress in this field, and Britain seemed to be losing the lead. Perhaps inevitably then, the wave-power issue remained firmly on the political agenda, with the ACORD assessment of it clearly still not being accepted by all. Indeed, that was true for the whole assessment process, which, as we shall see, was regularly revisited, as were divergent views of what should be done next.

3.2 The Select Committee Report and the Wave-Power Debate Revived

In 1984, the House of Commons Select Committee on Energy reported in fairly critical terms on the energy R&D programme, focusing its attack in particular on the DEn's handling of the wave-power programme:

> The evidence on the wave energy programme hardly inspires confidence in the Department's sponsorship of renewable energy R D & D, for the following reasons:
>
> (a) Despite it being obvious to all concerned from the outset that not much of value could be learned without sea trials on a reasonable scale, none have taken place:
>
> (b) The Department's enthusiasm for wave energy was apparently beginning to wane in the early 1980s, but the situation changed suddenly

in early 1982. As ACORD's deliberations remain secret, we merely point to the fact that there was a new Secretary of State (Mr Nigel Lawson) who was known to be eager to reduce public expenditure, that two known supporters of wave energy (Sir Herman Bondi and Dr F Clarke) had recently left the scene, and that there was a new Chief Scientist (Dr Challis).

(c) In slashing the wave energy programme, the Department – and, by implication, ACORD – allegedly used the argument that wave energy could not meet a specific cost target, when neither they nor anyone else was in a position to know whether this was the case or not. The whole reason for embarking on the programme was to create an option for a future in which energy prices might be considerably higher than today (the 'insurance' argument.) There is no evidence that the Department has ever assessed the much larger expenditures on the fast breeder reactor and fusion research against a similarly stringent cost criterion – certainly not at such an early stage in their development.

(d) It has been the Department's avowed aim to be in a position to decide in 1985 which renewable energy options to carry forward to prototype development. Wave energy, previously the Department's favourite horse, was effectively withdrawn before the race began. In consequence, we doubt whether a competent project team could now be re-assembled without great difficulty to pursue the work vigorously from where it was left in 1982.

The Committee was clearly unhappy with progress, especially on wave power (see Box. 3.1). It concluded: 'If there was any truth in earlier assertions that with wave power Britain was leading the world in a new technology with exciting prospects, that claim now has a hollow ring. We withhold judgement about the Department's current enthusiasm for wind energy, but note that the Department and ETSU (and, by implication, ACORD) attached the lowest priority to wind energy not many years ago. We are driven to the same view as the former Select Committee on Science and Technology – that the renewables R D & D programme has not been carried forward with the commitment and consistency which the objectives required. Ten years have slipped by since the 1973–74 "oil shock" and there is little progress to show as far as renewable energy sources are concerned' (Select Committee on Energy 1984).

Box 3.1 Select Committee Analysis of the Wave Energy Programme

The Committee noted that 'when the Department of Energy's wave-energy programme began in 1976 the objective was to identify the size of the resource through a series of supporting studies and R & D on four competing systems – the Salter Duck, the Cockerell raft, the Russell Rectifier and the NEL Oscillating Water Column. The emphasis was on paper and laboratory feasibility studies, but all concerned knew that real knowledge could be gained only from sea trials. The object from the outset, therefore, was to select the most promising devices and test reasonable scale versions at sea. The programme has strong support from two Chief Scientists (Sir Walter Marshall and Sir Herman Bondi) and the chairman of the Wave Energy Steering Committee (Dr F. Clarke). Consulting engineers were retained to improve knowledge of device efficiency and capital and operating costs.'

It is perhaps worth noting at this point that, as Wilson reports, Walter Marshall, who was DEn's Chief Scientist until 1977, while also being Harwell's Director, and from 1981, head of the UKAEA and then (from 1983) the CEGB, was indeed supportive of wave energy, despite his obvious nuclear sympathies (Wilson 2010). But at the time of the ACORD review, as the Committee noted, a new Chief Scientist was in post, Dr Anthony Challis.

The Select Committee went on ACORD reviewed the programme in 1978 and 1979 and advised the programme team of the importance which the Council (ACORD) attached to a sea trial near full-scale. The next objective for the programme team was to reduce the generation costs to 10p/kWh or below. In June 1980 they were asked by ACORD to select the most promising devices and to identify one to be taken to sea trials. In March 1981, ACORD set the team a deadline of March 1982 to select a prototype device and to halve the cost target from 10p to 5 p/kW. In anticipation of the 19 March 1982 meeting of ACORD, the programme team prepared a further set of cost estimates and recommended a further 12 months work to finalise details of the prototype for sea trials. Unlike all previous ACORD meetings at which renewables R, D & D had been discussed, the programme managers were excluded (by the Chief Scientist) from this meeting. At that stage, the device teams were close to the 5p/kWh target and some thought they could achieve a lower figure. However, at around the same time, ETSU had advised ACORD that unless wave energy could achieve under 4p/kWh it 'did not match in with what the system needed'.

It concluded 'The Department issued a Press Release on 27 April 1982 formally announcing its decision, on the advice of ACORD, sharply to reduce funding for the wave-energy programme, although ETSU's analysis (on which ACORD's recommendation was allegedly based) was not released until December 1982. The result was that £15 million had been spent abortively with the object of undertaking sea trials near full-scale, the projects then considered to hold the greatest promise were run down, the sea trials were not undertaken, and the only remaining parts of the wave energy programme were two devices previously considered to be the higher cost ones and which were still at the small scale laboratory stage' (Select Committee on Energy 1984).

Clearly the Select Committee were less than happy with the way the review had been conducted, and implied that wave power could have been found to be viable, if given a reasonable chance: it 'was effectively withdrawn before the race began'. Certainly, at the start of the programme, while the prospects for wave energy were uncertain, all agreed that the resource was large, and some obviously felt that wave energy had been harshly treated, with the Department receiving bad advice. Although the uncertainties were admittedly large, several commentators pointed out that, when nuclear power was at a similar stage of development (in the 1950s), the unit cost uncertainties were similarly large and yet the nuclear programme was pursued with considerable vigour and money.

Nevertheless, the official view on wave power may turn out to have been at least partly correct. After all, the claims made by the device teams are likely to be somewhat more optimistic than those produced by independent assessors. Certainly wave energy has not moved ahead much subsequently in the UK or anywhere else, but that may have been at least in part due to the lack of support by governments, with the UK programme, being the most advanced and influential, also setting the tone elsewhere. We will be coming back to look at the strategic issues raised by the handling of wave energy in the Conclusions chapter and, as we shall see, the issues continued to be debated throughout the period covered by this book.

Back in immediate post-ACORD review period, the UK wave-power device teams had been put in a difficult position. Some commentators have argued that part of the problem was the undue sense of urgency that prevailed: the rival device teams evidently felt they were in a race to get cost estimates down sufficiently to justify continued funding. They were obviously right! But this did not necessarily lead to good engineering.

After all, there was, in reality, no immediate urgency: theoretical and tank studies could have been spread over several years before going to sea with prototypes. However, testing a working prototype at sea obviously had attractions in terms of political expediency.

Time pressures like this are obviously just some of the problems facing innovators working, inevitably, in an uncertain technical and financial environment. But the researchers felt their problems were not eased by the way the assessment had been performed, or by the way the programme had been managed. As one researcher, Stephen Salter, commented, somewhat bitterly, before the ACORD decision: 'I am sure that many people engaged in the work whether from device teams, consultants or administration will agree with me when I point out that improvements could be made to the administrative machinery. So much of it seems designed to delay and demoralise' (Salter 1981). Another participant had put his criticisms of the wave energy steering group's activities even more bluntly, suggesting that the researchers were often kept 'in the dark' so that they 'did not know which way they were steered' (Flood 1982).

Of course one might expect bitter criticisms like this from hard-pressed researchers. It seems that innovators are nearly always faced with such problems. No doubt similar complaints could be collected from, for example, the wind-power teams, even given the fact that, following the 1982 ACORD review, on-land wind power became increasingly favoured. We will be returning to look at issues like this later, in our retrospective review of the UK renewables programme.

3.3 Tidal Power and the Severn Barrage

As noted earlier, tidal barrages were reviewed separately and were not included in the ACORD review, with an overall positive report emerging on them in 1981 from a team led by Sir Herman Bondi (Bondi 1981). To some extent, barrages were therefore insulated from the at times rather acrimonious debate surrounding the ACORD review. In part, their exclusion was perhaps since, unlike the other renewables, they were seen as less in need of technical R&D effort. However, there was clearly a need for research on likely environmental and social impacts, issues that had moved up the agenda, given the opposition to large barrages by some key environmental groups and concerns about possible objections to these large potentially very invasive structures from the public (Barac et al. 1983).

A series of environmental studies were carried out, with the DEn subsequently publishing detailed studies by ETSU (jointly with industrial consultants) of the environmental impacts (ETSU 1989a, b). There was less opposition to smaller barrages, and ETSU also looked at the proposed Mersey Barrage (ETSU 1988), as well as some others. See Box 3.2. But it was the large 8.6 GW Severn Barrage that drew most attention, with the DEn producing a major report on it in 1989 (Department of Energy 1989). It was generally supportive, but did point to the problems of funding such a large project.

Box 3.2 Tidal Barrages and Other Tidal Options in the UK

The Tidal Barrage idea has a long history, with one of the first proposals for a barrage on the Severn Estuary (although not including power production) dating back to 1849. The first indication of interest in power production came in 1920, from the UK Ministry of Transport, and this was followed up in a study started in 1925 by the Brabazon Commission. Although its proposal for an 804 MW barrage did not get pursued, the idea clearly continued to excite interest over the years (Calder 1938).

The first major contemporary study came in 1977 (Department of Energy 1977), with interest also emerging in sites other than the Severn. In theory, the total UK exploitable potential, assuming that every practical UK barrage scheme was developed, was put around 53 TWh p.a., or about 14% of UK electricity generation. Given that the Severn had one of the largest tidal ranges and could accommodate a large barrage, a barrage there would make the largest contribution, approximately 17 TWh p.a., from an 8.6 GW installation, but estimates suggested that the Wash, the Mersey, the Solway Firth, Morecambe Bay and possibly the Humber could also make significant contributions. Studies were carried out of each. In addition to these medium-scale sites, there were many smaller estuaries and rivers that could be used. Feasibility studies were carried out on the Loughor Estuary (8 MW) and Conwy Estuary (33 MW) in Wales, the Wyre (64 MW) in Lancashire, and the largest, Duddon (100 MW), in Cumbria (ETSU 1989c). Overall, the total UK potential for small tidal barrage schemes (i.e. schemes of up to 300 MW capacity) was estimated at 2% of electricity requirements, with the most economic of these providing a total of 1.5% of UK power requirements.

Some interest had also been shown in small tidal current devices, using the horizontal tidal flows rather than the vertical rise and fall of the tides, as with so-called tidal range system like barrages. That would avoid having to build invasive barrages (Peachy and Wyman 1979; Fraenkel and Musgrove 1979). As we shall see, subsequently this idea was to receive much more attention. So did the idea of tidal lagoons, since, unlike barrages, these free-standing impoundment structures would not block off entire estuaries, but would still work on the tidal range.

Much of the running on the economic side had been made by the Severn Tidal Power Group, the industrial consortium led by McAlpines. It estimated that, depending on the price at which the electricity could be sold, the internal rates of return on the large capital outlay for the Severn Barrage, at that time put at around £10 billion, would be around 6–8% (STPG 1986). That was seen as unlikely to be attractive to private sector investors, who might expect much higher rates of return, up to 10% or perhaps more. So it would have to be publicly financed. Given that, once built, the barrage could run for at least 100 years without major extra costs, public finance seemed reasonable. Although that was not forthcoming, interest in barrages, large and small, remained, with, as we shall see, tidal technology generally being henceforth included in subsequent assessments of renewables, rather than being seen as falling in a separate category. However, although on the agenda, barrages were still mostly seen as large national projects, which were some way off, with no decisions on major funding being needed at this point, whereas, in terms of immediate practical developments, it was wind power, and the other renewables that had survived the ACORD review, that led the way.

3.4 The Continuing Policy Debate and a New ACORD Review

In the post-ACORD review period, the debate moved on to what should be the appropriate level of funding as the next phase got underway. Speaking to the all-party Energy Studies Group in October 1983, Dr Challis, who had then just retired as Chief Scientist at the DEn, pointed out that, after the (then) current phase of work was completed, the successor programme, aiming to terminate around the early stages of commercialisation, 'would need to be funded... at between £100 and £200 million each year' (Challis 1983).

Dr Challis did not suggest that *all* the money should be found by government, 'but the timescale risk and longer term profit potential at this second phase is such that major state funding would have to be found to lead in private money'. A similar point was made by Lord Ezra in the House of Lords in May 1984. He argued that the second phase, which he saw as starting in 1985, should be allocated £100–200 million.

However, in November 1984, the Parliamentary Under Secretary of State for Energy, David Hunt, made clear that the £14 million per annum public expenditure that renewables were by then enjoying was likely to remain unchanged 'for the next few years' (Hunt 1984). In effect, extra funding was to be frozen. Then, in January 1985, Mr Hunt spelt out the official view concerning the overall renewable programme in more detail. 'Renewable energy sources in the UK do offer the prospect of eventual diversification in energy supplies. However, none can substitute for a main-line fuel in the short or medium term. These sources should make a valuable and growing contribution in the next century as they become increasingly competitive.'

He went on 'Some renewable energy technologies, for example the combustion of certain waste materials and good passive solar design in buildings, are economic now. Others, like on-shore wind turbines, have the promise of being so in the medium term, with geothermal hot dry rocks perhaps becoming cost-effective in the more distant future. Our approach, in the case of those technologies which look commercially attractive now, is to provide support for demonstration and other appropriate aids to make the users aware of the opportunities. For those technologies where we do not yet know whether costs can be brought down to competitive levels, we are supporting research work aimed at reducing costs and improving performance, knowledge and understanding.'

He concluded 'It is essential in all research and development, as programmes mature and results accumulate, from time to time to re-assess priorities and redefine objectives. The Department is now engaged in a major review of its renewable energy activities, similar to that carried out in 1982, the main aim being to establish a strategic framework for the remainder of the decade' (Hunt 1985a).

Recommendations resulting from the new review which Hunt referred to, carried out by ACORD and the DEn, emerged in June 1985, and a government response was made in July. There were few surprises. On-land wind, tidal and deep geothermal were seen as 'promising'. Some biofuels and passive solar were 'attractive', active solar, offshore wind and photovoltaics were 'long shots', while wave power was finally rejected, along with geothermal aquifers.

The new ACORD recommendations were based on ETSU's 'R30' overview report, 'Prospects for the exploration of the renewable energy technologies in the United Kingdom', produced by the Chief Scientist's Group. That included ETSU's revision of estimates for *technical potential,* which R30 explained 'takes account of the amount of energy which is, or could be made, available from the resource, judging where appropriate, the future state of the technologies'. But ETSU added that it also took into account 'preliminary judgments concerning environmental acceptability'. It also looked at the potential for rapid rate of development (in the case of emergencies i.e. the 'insurance' role) and the export potential.

Overall though, it noted that the technical potential was still a key criterion 'which shaped the initial renewable energy programme when development of the technologies was barely advanced'. Interestingly, only biofuels (energy crops from forests/farms) were forecast to make a major contribution in 2025, along with a small contribution from on-land wind, the tidal and hydroelectric contributions being unclear (ETSU 1985a).

ETSU's *RE News* (October 1985), reporting on these assessments, commented that 'naturally there are uncertainties in applying some of these criteria in a quantitative way, and so the analysis involved considerable application of judgement but special attention was given to apply that judgement uniformly across the technologies'.

Under the classification system used, each technology was assigned to one of three broad categories:

- Economically attractive,
- Promising, but uncertain,
- Long shots.

The definition and interpretations of the classifications are spelt out in Box 3.3.

Box 3.3 R30's Classification System

R30 says that '**Economically attractive** means that a technology seems likely to make a significant and cost-effective contribution to UK energy supplies whether or not the cost and performance of the technology can be improved and whether or not fuel prices rise in real terms in the near future. (...) Waste combustion and passive solar design are examples in this classification. They are far enough advanced to be costed accurately and are competitive against conventional fuels at current prices. They do however require demonstration before the market-related barriers to their uptake can be overcome and they become available for widespread application.

Promising but uncertain means that a technology is likely to become competitive if predicted costs can be achieved or if fuel prices rise in real terms. Evaluation of the likely impact of such technologies requires an understanding of the uncertainties arising both in the evaluation of the impact of R & D on the performance of the technology and in the projected future fuel prices. Wind turbine generators and geothermal hot dry rock energy systems provide examples in this category though on different time scales. These technologies are currently being advanced by large R & D projects and much development and demonstration work remains to be done.

Long shots are those technologies which might be deployed cost-effectively only in the unlikely event of a dramatic improvement in the costs and performance or if fuel prices increase sharply. Even then exploitation may occur only if the more attractive renewable energy technologies or other "insurance" technologies were not available. The evaluation of "long shots" requires not only the treatment of the dual uncertainty about R & D and about future fuel prices but also the treatment of extreme views of the future. Wave-energy converters and active solar space heating are examples of "long shots". Under the future fuel price assumptions of the Department of Energy's 1982 Energy Projections, these technologies would be expected on economic grounds to make no significant contribution to UK energy supplies in the foreseeable future. A technical breakthrough could change this position.'

R30 commented that 'The process of assessment and reassessment which underpins the above classification of the renewable technologies is continuous. The present analysis is largely an extension of previous assessments and offers a further snapshot recognising that different elements in the analysis are becoming clearer at different times as R & D progresses' (ETSU 1985a).

Subsequently, ETSU's *RE News* (October 1985) added that 'ACORD considered that the "long shot" category should be further divided into those technologies where benefit might be obtained, from maintaining a minimum level of R & D with the necessary international collaboration, i.e. the real "long shots", and into those where effort should be limited simply to keeping up to date with progress elsewhere and on which, otherwise, expenditure should cease completely i.e. "the rejects".'

ACORD recommended that wave power should be relegated to this latter category and that the current geothermal aquifer programme should be wound up. In a statement in the House of Commons in July 1985, in response to a parliamentary question, David Hunt confirmed that he had 'accepted ACORD's advice to discontinue work on wave power and geothermal aquifers in order to concentrate resources on more promising technologies', essentially wind, geothermal hot rocks, passive solar and biofuels, while he awaited the outcome of studies 'into the private financing of the Severn Tidal Power Scheme' (Hunt 1985b). See Box 3.4. He also reported that 'the estimated expenditure by my Department on renewable energy research and development in 1985–86 is £14 million'. So there would be no overall increase on the 1984–1985 allocation.

Box 3.4 Hunt's Decisions in Response to the New ACORD Review

Outlining the resultant decisions from the new review, Parliamentary Under Secretary of State for Energy David Hunt said 'The major programmes on wind power and geothermal hot dry rocks will continue. They will be further reviewed when experience is available from the wind-powered generators under construction on Orkney and at Carmarthen bay, and when the present phase of the geothermal experiment at Rosemanowes in Cornwall is complete. I agree that our present programme of research and development into passive solar design of buildings is important and justified. I have been glad to note ACORD's recognition of its economic attractiveness, and the need to involve the private sector in its application. The biofuels programme will continue, and this reflects especially the economic potential of waste as a fuel. Support for small-scale hydro-power technology will also continue. Studies into the private financing of the Severn tidal power scheme continue, and I await their outcome. I have accepted ACORD's advice to discontinue work on wave power and geothermal aquifers in order to concentrate resources into more promising technologies. However, on wave power in particular, I recognise that it will be important for the Department to be receptive to new ideas which might succeed in achieving the major cost reductions necessary to make it a worthwhile source of power' (Hunt 1985b).

In the linked DEn press release, he explained that 'the thrust of our programme is to back the winners. We are concentrating our resources on the most promising options.' As an example, he cited passive solar: 'We are doubling our support for passive solar – the direct solar heating of buildings – which is a potential winner and makes good economic sense. We shall be encouraging the private sector to design buildings which make the most use of sunshine because it can cut heating and lighting costs by as much as 30%. Solar houses are economical and people who live in them like them.'

Passive solar has the attraction of not requiring expensive (active) technology or government support, just a change in building design, with better insulation and optimisation of solar gain. However, interestingly, the detailed research actually showed that the solar gain element only offered a small part of the overall energy saving in UK conditions: most was due to the improved insulation (Chapman et al. 1985). So that was the real winner. As we shall see, the passive solar programme was subsequently wound up, and although the concept (the daylighting aspects in particular) remains valid, it has mostly just become part of good building practice.

Overall, this second review created somewhat less of a public debate than the first, although the 'final' rejection of wave power did generate considerable disquiet within the renewable energy research community, especially given the fact that the Norwegians at this point were commissioning a small (500 kW) shore-based wave-power unit, with estimated unit costs of only 3.4 p/kWh, compared to the 9–15 p/kWh estimate, albeit for large deep-sea systems, in R30. This latter estimate was actually much higher than the 4–12 p/kWh given in R13, ETSU's earlier 'Strategic review of the renewable energy technologies' (ETSU 1982). In R30, however, ETSU commented that 'the probability of achieving a cost lower than 9p/kWh is small' and added 'although costs of 5–6p/kWh have been claimed for some devices, so far these are unsubstantiated' (ETSU 1985a).

3.5 An End of Term Review

A half-day debate on alternative sources of energy was held in October 1985 in the House of Commons (Hansard 1985). In effect, it provided a review of the state of play at that point. The Parliamentary Under Secretary of State for Energy, David Hunt, opened with a review of the Government's position and of the technical progress so far. He reiterated the Government's view that 'market forces are the best way of discovering exactly what customers want and the best means of finding out whether they can have it at a reasonable price'. He added: 'We have been concerned to set a framework which will ensure that the markets operate in the energy sector with a minimum of distortion and that energy is produced and used efficiently.'

On renewables, he reviewed ETSU's R30 conclusions and the various programmes, and then stressed that, while he 'believed that renewable alternative sources of energy are a key element in UK energy strategy … our policy is to get the best possible value for money from our research and development funds and back the winners'.

That last phrase was criticised by several of the members who spoke subsequently in the debate. Alex Eadie (Labour) commented that 'backing only winners in research and development is a contradiction in terms… There will be losers in research and development.' He called instead for 'technological pluralism' and recommended a dramatic increase in R&D funding for renewables, citing a programme involving the allocation of £150 million per annum.

The conservative MP, Tony Speller, then chairman of the Parliamentary Alternative Energy Group, was also concerned: 'Sooner or later R & D requires a working size plant so that things are seen to work… We could do so much with a little added to the pennies that we already spend.' More aggressively, Malcolm Bruce, from the Liberal-SDP Alliance, invoked a conspiracy theory. 'It seems that there is a conspiracy afoot to prevent the full-scale development of alternative sources of energy.'

In all there were 12 contributions from all sides of the House, many of them sharply critical of the Government's handling of the wave-power decision. According to Eadie: 'the decision was a blunder and the Government are in danger of making themselves look foolish'. Bruce commented: 'Those who have received representations from people involved in the wave programme will have been struck by the degree of bitterness with which they have greeted the cut in that programme, not to mention the bitterness they feel about the way in which they are judged by comparison with the conventional and established forms of energy renewal, both in terms of expected rates of return and the amount of investment that they are allowed to use to overcome their problems. Just when they think that they are getting there, the funding is cut from under them and the opportunity to go ahead with something with real potential is closed off.'

Moreover, Speller insisted that 'we must return and consider the Norwegian experiment, which uses British technology. It is too easy to say that it is a long shot and we should forget it. An island nation which has tidal water flowing to and fro, hydropower and other sources of water should not write off wave power as a long shot. There is fantastic power available.'

In his concluding response, David Hunt defended the Government's decision on wave power, and asserted that, while renewables 'are unlikely to make a significant contribution to UK electricity supplies this century… they should make a contribution in the twenty-first century which will grow as they become increasingly competitive'.

Perhaps the last word on this phase of the UK renewables programme should go to the then Secretary of State, Peter Walker, who, at a meeting of the Parliamentary Alternative Energy Group in 1985, commented, in response to a question implying that renewable R&D had been cut back, that far from there having been cuts, given the scale of the public expenditure cuts in recent years generally, renewables had 'come out quite well' (Walker 1985).

Standing back from the flurry of events and policy conflicts reported above, it may be worth pointing out that the sums of money being contested were relatively small, tens of millions, at a time when very much more was being spent by the government on nuclear power, typically hundreds of millions per annum for R&D. Table 3.1 shows the spending on renewables in the period 1979–1984, and an estimate for 1984–1985. The Departmental allocation for nuclear R&D is also shown. However, in addition to the very much larger sums for new nuclear plant construction, there was also non-departmental R&D funding, with total public sector funded nuclear fission R&D, although falling from a high of around £460 million in the 1970s, still running at £300–350 million in the early to mid-1980s, before falling to £150–200 million in the later 1980s (Science and Technology Committee 2011).

The priorities did seem skewed and at odds with public views on nuclear power. Following the Three Mile Island accident in the USA in 1979, support for nuclear power in the UK dropped notably, and it fell even further after the Chernobyl accident in April 1986, with, by May

Table 3.1 Funding for renewables 1978–1985

	1978–1979	1979–1980	1980–1981	1981–1982	1982–1983	1983–1984	1984–1985 estimate
Wind	0.3	0.6	0.8	0.9	2.5	2.5	4.8
Wave	1.8	3.0	3.3	4.4	3.1	1.0	0.4
Geothermal							
Aquifers	0.2	1.3	1.7	2.6	1.5	1.8	1.1
Hot Dry Rock	–	–	0.6	5.6	3.0	2.9	4.6
Solar	0.2	1.2	0.9	0.7	1.4	0.9	0.5
Biomass	–	0.1	0.3	0.4	0.7	0.3	0.2
Tide	–	0.6	1.4	0.4	–	–	0.3
ETSU Services	1.1	1.4	2.2	2.3	2.0	1.9	2.1
Total renewables	3.6	8.2	11.2	17.3	14.2	11.3	14.0
Nuclear R&D	not given	146	188	206	216	206	198

Department of Energy expenditure in £m on renewable energy research and development (ETSU 1985b) Nuclear data (Hansard 1984)

1986, only around 18% of those asked in a Gallup poll supporting an increase in nuclear power, as against 39% calling for 'no more at present' and 36% saying 'stop nuclear power'. Support for new nuclear continued to fall, to around 13% in 1991, while 78% of those asked either wanted 'no more nuclear plants at present' or for the use of nuclear power to be halted (Gallup 1994; Elliott 1997). By contrast, although, in the early days, the prospects for renewables were less clear or well known by the general public, support was strong: 87% of the respondents to a Gallup poll in the UK, carried out in 1991 for Friends of the Earth, indicated that they would prefer the government to increase spending on renewables. We will be looking at subsequent opinion trends later on. Although opposition to nuclear reduced somewhat, possibly due to increased concerns over climate change, the disparity remained basically the same, with renewables being strongly preferred (Ipsos MORI 1998).

Public opinion on the development of energy alternatives had no doubt been shaped in the 1970s by the sometimes alarming experiences with the oil crisis and then by the Three Mile Island and Chernobyl nuclear disasters: changes were clearly needed. Given that backdrop, the new renewable technologies captured more media attention. That included wave power, with, amongst others, wave energy pioneer Professor Stephen Salter proving to be particularly telegenic. Despite the wave energy programme being marginalised, the intrigues and policy reversals in relation to it received media coverage, and, perhaps perversely, that may have raised public awareness that there was a range of interesting new options. This sort of high-tech hardware was certainly some way on from the 'hippy' imagine that 'AT' sometimes had.

Meanwhile, environment pressure groups continued to campaign for more attention to be given to the full range of renewables (Flood 1983), usually in the context of them being a clear alternative to nuclear power (Boyle 1988). In parallel, NATTA, the independent Network for Alternative Technology and Technology Assessment, which I had helped to set up in 1976, continued to promote renewables via a series of conferences around the UK, as well as its newsletter, launched in 1979. In the 1980s, NATTA focused mainly on community-scale technology, but also sought to subject larger-scale renewable energy technologies like tidal barrages, and also medium-scale wind farms, to critical assessment (NATTA 1981a, b; Clarke 1988).

However, with privatisation on the agenda, as we shall see, the policy debate at government level was to be increasingly become focused not so much on technology, of whatever sort or scale, but on markets and major political changes to them, although that of course was to drive the next stage in the development of renewable technology.

References

Barac, C., Elliott, D., & Spencer, L. (1983). Public Awareness of Renewable Energy; a Pilot Study. *International Journal of Ambient Energy, 4*(4), 199–211.

Bondi, H. (1981). *Tidal Power from the Severn Estuary*, Energy Paper 46. London: Department of Energy.

Boyle, S. (1988). *An Alternative Energy Path for the UK*. London: Friends of the Earth.

Calder, R. (1938, January 29). Mighty Engines Driven by the Moon. *Modern Wonders*.

Challis, A. (1983). Comment at a Meeting of the All Party Parliamentary Group for Energy Studies (PGES), Oct. as Reported in NATTA Newsletter 11, Sept-Oct 1984. Full Text in Energy Focus Journal, Issue 1, Longmans.

Chapman, J., Lowe, R., & Everett, R. (1985). *The Pennyland Project*. Milton Keynes: Energy Research Group, The Open University.

Clarke, A. (1988). *Windfarm Location and Environmental Impact*. Network for Alternative Technology and Technology Assessment Report, Milton Keynes.

Department of Energy. (1977). *Tidal Power Barrages in the Severn Estuary*, Energy Paper 23. London: Department of Energy.

Department of Energy. (1989). *The Severn Barrage Project: General Report*, Energy Paper 57. London: Department of Energy.

Elliott, D. (1997). *Energy, Society and Environment*. London: Routledge. On line at: https://docslide.us/documents/energy-society-and-environment-routledge-introductions-to-environment.html

ETSU. (1982). *Strategic Review of Renewable Energy Technologies*. Oxfordshire: Energy Technology Support Unit, Department of Energy (R13).

ETSU. (1985a). *Prospects for the Exploration of the Renewable Energy Technologies in the United Kingdom*, Overview Report Produced by the Chief Scientist's

Group, R30. London: Energy Technology Support Unit, Department of Energy.
ETSU. (1985b, October). *RE News*. Harwell: Energy Technology Support Unit Newsletter.
ETSU. (1988). *Tidal Power from the River Mersey: A Feasibility Study Stage 1*, ETSU TID 4047. London: Mersey Barrage Company Limited/Energy Technology Support Unit, Department of Energy.
ETSU. (1989a). *Severn Barrage Project: Detailed Report, Vol. I., Tidal Hydrodynamics, Sediments, Water Quality, Land Drainage and Sea Defence*, TID 4060 p1. London: Severn Tidal Power Group/Energy Technology Support Unit, Department of Energy.
ETSU. (1989b). *Severn Barrage Project: Detailed Report, Vol. II., Ecological Studies, Landscape and Nature Conservation*, TID 4060 p4. London: Severn Tidal Power Group/Energy Technology Support Unit, Department of Energy.
ETSU. (1989c). *The UK Potential for Tidal Energy from Small Estuaries*, ETSU TID 4048 – PI. London: Energy Technology Support Unit, Department of Energy.
Flood, M. (1982). *Quoted in 'Eclipse of the Sun?'*. London: Friends of the Earth.
Flood, M. (1983). *Solar Prospects*. London: Friends of the Earth, Wildwood House.
Flood, M. (1985, June). The Troughs and Crests of Wave Energy. *CME Journal*.
Fraenkel, P., & Musgrove, P. (1979, January 30–February 1). Tidal and River Current Energy Systems. *Proceedings of the IEE Conference on Future Energy Concepts* (pp. 114–17). London: Institution of Electrical Engineers.
Gallup. (1994). Gallup Poll Data from the UK Governments Nuclear Review, Submission by Stop Hinkley Expansion, Sept. Full Data Presented in Elliott 1997 op cit, p. 31.
Hansard. (1984, July 25). Written Reply to a Parliamentary Question. *Hansard, 64*, 633W. https://api.parliament.uk/historic-hansard/written-answers/1984/jul/25/alternative-energy-sources
Hansard. (1985, October 25). Debate on Alternative Sources of Energy, House of Commons, Official Report Transcript. *Hansard, 84*, c. no. 169. https://api.parliament.uk/historic-hansard/commons/1985/oct/25/energy-alternative-source
Hunt, D. (1984). *Speaking to a PRASEG Meeting*. Parliamentary Renewable and Sustainable Energy Group, House of Commons, November, as Reported in NATTA Newsletter 33, Jan–Feb 1985.

Hunt, D. (1985a, January). *Interview in RE News, ETSU's Newsletter.* Harwell: Energy Technology Support Unit.
Hunt, D. (1985b, July 2). Parliamentary Answer, House of Commons. *Hansard, 82,* cc85-6W. https://api.parliament.uk/historic-hansard/written-answers/1985/jul/02/renewable-energy-sources#S6CV0082P0_19850702_CWA_4
Ipsos MORI. (1998). *Government Energy Policy Swims Against the Tide of Public Opinion.* Ipsos MORI Poll in 1998. https://ems.ipsos-mori.com/research-publications/researcharchive/2002/Government-Energy-Policy-Swims-Against-the-Tide-of-Public-Opinion.aspx
NATTA. (1981a). *Community Action and Alternative Technology.* Network for Alternative Technology and Technology Assessment Report, Milton Keynes.
NATTA. (1981b). *Alternative Technology: An Assessment of Technical, Environmental and Institutional Problems.* Network for Alternative Technology and Technology Assessment Report, Milton Keynes.
Peachy, C., & Wyman, P. (1979, January 30–February1). Tidal Current Energy Conversion. *Proceedings of the IEE Conference on Future Energy Concepts.* London: Institution of Electrical Engineers.
Salter, S. (1981, April). Wave Energy Costs' in 'Economic Costs of Alternative Energy Sources'. *Proceedings of a PARLIGAES/Energy Manager Conference.* London: Parliamentary Liaison Group for Alternative Energy Strategies.
Science and Technology Committee. (2011). *Nuclear Research and Development Capabilities.* House of Lords Science and Technology Committee, Third Report, Chapter 2, Para 11, Figure 1. https://publications.parliament.uk/pa/ld201012/ldselect/ldsctech/221/22105.htm#note8
Select Committee on Energy. (1984). *Energy R, D&D in the UK.* House of Commons Select Committee on Energy, Session 1983–84, Ninth Report.
STPG. (1986). *Tidal Power from the Severn.* London: Severn Tidal Power Group.
Walker, P. (1985, October 9). *Comment at a Parliamentary Renewable and Sustainable Energy Group Meeting.* House of Commons.
Wilson, J. C. (2010). *A History of the UK Renewable Energy Programme, 1974–88: Some Social, Political, and Economic Aspects* (PhD Thesis). University of Glasgow. http://theses.gla.ac.uk/3121/1/2010WilsonJohnPhD.pdf

4

Forward to the Market *into the 1990s*

4.1 Restructuring the Electricity Supply Industry

Following the return of another Conservative Government to power in 1987, the next main focus in the energy field was the debate over the proposed restructuring of the UK's electricity supply industry. British Gas had been privatised in 1986, and there were plans for the electricity industry to follow.

The emphasis was very much on enhancing competitive markets. This was reflected in a response, in January 1988, from Michael Spicer, then an Energy Minister, to a Parliamentary Question on the role of subsidies and grants. Spicer commented 'the renewable technologies would not be best served in the long term by distorting the market by grant aid or other subsidies for their use' (Spicer 1988). While the government would continue to support some R&D, beyond that, it was now up to the private sector, with privatisation bringing about some major changes. Some, like Spicer, thought renewables might benefit, as a free energy market was established, but for many, the prospects of having to face market pressures outside of the protection of the nationalised system was challenging.

D. Elliott, *Renewable Energy in the UK*, Energy, Climate and the Environment,
https://doi.org/10.1007/978-3-030-04765-8_4

In 1988, as something of a last gasp, the soon to be privatised CEGB decided to launch a publicly funded wind-power programme and announced that it would build three 'wind parks' to assess the viability and acceptability of wind power in the UK context. In parallel, there was also much discussion of the merits or otherwise of the Severn Tidal Barrage. However, as the debate over privatisation moved on, it became clear that major projects like this were unlikely to be backed. Equally, it also began to be clear that nuclear power might also face problems in the new privatised market.

The debate over privatisation ran in parallel with a Public Inquiry, which started in 1988, and ran into 1989, over a proposed new nuclear plant at Hinkley Point in Somerset. Amongst other things, the Inquiry provided a forum for discussion of the prospects of renewables as an alternative policy option, as well of course of the issue of nuclear power's fate under privatisation (Roberts et al. 1991). In the event, although the Inquiry decided in favour of the Hinkley project, in the post-privatisation context, it did not go ahead. Almost 30 years on, a new Hinkley project is now going ahead, but as we shall see, the level of private and public financial support it needs is controversial.

Back in the 1980s, privatisation clearly also presented some of the larger renewable energy projects, like tidal barrages, with problems. Nevertheless, the prospects for some of the smaller-scale more modular renewables could be quite bright, as the rapid growth of wind power in the USA and Denmark showed: see Box 4.1. With privatisation looming, and the experience of the USA and Denmark in mind, the emphasis within the renewable energy debate therefore shifted even further towards the prospects for commercialisation.

Box 4.1 Wind-Power Developments

The USA had been pushing ahead with wind power quite seriously from the mid-1970s onwards and had adopted a 'high tech' aerospace approach, with, as noted earlier, the emphasis on increasingly large complex prototypes, like the 2 MW Boeing/NASA 'Mod 2' series. Germany and Sweden also embarked on similar projects.

When the UK started up its own wind programme, in the early 1980s, as noted above, it adopted a similar approach. After building a smaller 250 kW prototype as a stepping stone, a large 3 MW machine was built by the Wind Energy Group (WEG) on the Orkneys. However, the large machines were not successful: there were some technical failures due, for example, to the large stresses on the giant blades, and in general, after the initial enthusiasm, they were seen as too big, complex and expensive, pushing the technology too far too soon. The big machines were abandoned, the 3 MW WEG unit, which had developed blade cracks, was down-rated and sold off and later demolished.

Around the world, commercial emphasis shifted to smaller machines of around 300–400 kW rated power. Denmark had already taken a lead by initially emphasising relatively simple, robust, machines, based on proven agricultural engineering approaches. As noted earlier, local agricultural engineers, operating on an almost craft basis, but with carefully targeted state support, developed small wind turbines which subsequently proved to be world-beaters. The Danish machines sold in great numbers to the USA and later to the UK and were then gradually scaled up as technical experience was gained and markets extended. Denmark, and then Germany, led the pack, with the UK buying in machines. The UK's first wind farm was at Delabole in Cornwall, which opened in 1991, using 10 Vestas 400 kW wind turbines from Denmark.

Nearly all these machines (the large two-bladed WEG and Mod 2 series apart) were three-bladed horizontal axis propeller-type devices. There was initially some interest in vertical axis designs, since, in that format, the generator unit could be on the ground rather than at the top of the tower, and the turbine would accept wind from any direction rather than having to be swung to face it. As noted earlier, the UK tested a novel 'arrow-head' vertical axis machine at Carmarthen Bay. However, vertical axis wind designs were not followed up around the world, in part since, with new materials, it was easier to extend conventional rotor lengths, to increase the area swept, although the inherently more stable vertical axis configuration has recently (in the 2010s) been looked at again in the context of floating offshore wind devices.

In 1988, the DEn published a major new report, Energy Paper 55, *Renewable Energy in the UK: The Way Forward*, which summarised their overall renewable energy strategy in terms of three basic phases (see Box 4.2), the first being essentially the 'identification of potentials' phase, the next two involving a move to commercial projects. This was to be the main focus of the next stage. Energy Paper 55 also mapped out the funding plan for the next phase: it suggested that the DEn's R&D budget would peak at

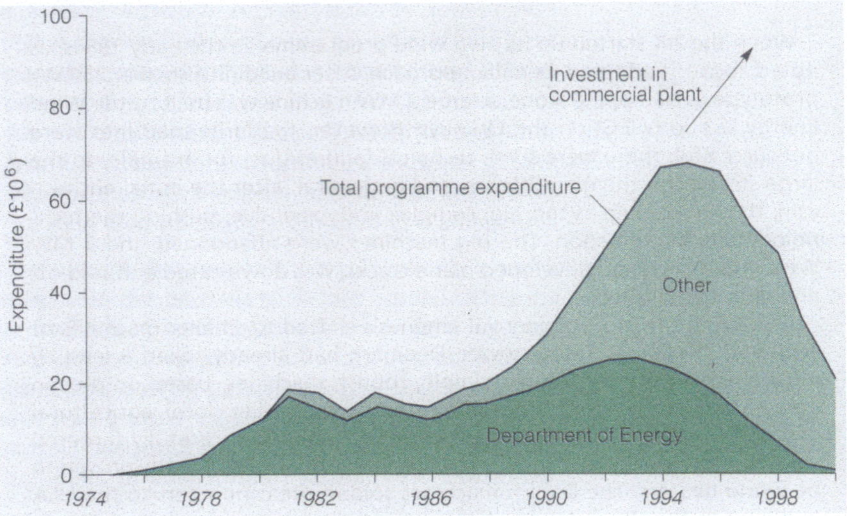

Fig. 4.1 EP 55: DEn's expected R&D funding pattern. (Department of Energy 1988)

around 1993, and then fall to almost zero by 2000 (see Fig. 4.1), with the private sector expected to take over any promising lines of development, and invest in relevant full-scale commercial projects.

Box 4.2 Renewable Energy in the UK: EP 55 (Department of Energy 1988)

The programme can be perceived in three phases:

- *Phase 1—Mid-1970s to mid-1980s*

The programme identified feasible technologies, their costs, their technical potential as contributors to UK energy supply and the possible time scale of that contribution. This phase was conducted mainly by analysis and experiment without heavy financial commitment to technology development at full scale. Further expenditure on those technologies unlikely to contribute significantly in the UK was curtailed and increased effort was focused on the more promising ones.

- *Phase 2—Mid-1980s to early 1990s*

A collaborative, cost-shared technology development programme [was] conducted with industry in order to identify the contributions to be expected from those renewable sources which could be economic now in some market sectors or show promise of being so in the future. Also being developed and applied in this phase [were] the technology transfer processes necessary for the creation of a successful market.

- *Phase 3—Early 1990s to late 1990s*

It is envisaged that in this phase of the programme the demonstration and technology transfer activities will be undertaken. The Department of Energy's part in the development of these technologies will thus be concluded by the end of Phase 3 and they will prosper or not according to commercial forces.

Evidently, the DEn was clearing the decks for privatisation, by mapping out ways in which R&D could be passed over to the private sector. In this context, the continuing debates over wave power no doubt were something of an embarrassment. Following continuing pressure for a rethink on wave power, in 1989, a new review was initiated, carried out by Tom Thorpe at ETSU. The Department did admit to some errors in the earlier assessments. For example, incorrect load factors were evidently used for Salters Duck, almost doubling the power cost estimate (New Scientist 1990), and other irregularities were alleged to have occurred (Ross 1990), although the new review was not designed to go over old ground, but rather to look afresh at the issues.

In 1991, the DEn also initiated a new overview study of its renewable energy strategy and set up a special Renewable Energy Advisory Committee (REAC) to feed in ideas. REAC was chaired by Energy Minister Colin Moynihan, who had taken a particular interest in renewables, and indeed seemed to be something of an enthusiast. As we shall see, these two reviews reported back in 1992.

However, by that time, following the re-election of a Conservative Government for a fourth term, the political framework had changed. With privatisation now complete, the DEn was abolished in 1992, the bulk of its work being absorbed into the Department of Trade and Industry (DTI). The Select Committee on Energy was also therefore wound up, but not before issuing a final very critical report, in which it commented 'it is difficult to regard the history of renewable R & D funding in the UK as other than a history of volte faces, premature judgements and plain errors' (Select Committee on Energy 1992). Clearly, it was still not impressed by the government's handling of renewables, but post-privatisation and the Committees' demise, the focus moved on.

Moynihan had lost his seat at the 1992 election, but the REAC review went ahead, and its report emerged at the end of 1992. It was very positive in general, suggesting that renewables could in principle supply more than twice the UK electricity requirements. More conservatively, in practice, it estimated that the UK could obtain around 20% of its electricity from renewable sources at competitive prices by 2025. Even so that was a significant amount. However, it was less optimistic about the prospects for some of the technologies. On wave power, it confirmed the view that further work should be abandoned. Similarly, work on aquifers, and even HDR geothermal, was seen as problematic. Work on vertical axis wind turbines was also seen as having failed to deliver (REAC 1992).

The wave-power conclusions were backed up by the Thorpe report (R72), which emerged at the same time. Thorpe had done a very comprehensive study involving all the key groups, but had concluded that the prospects for wave power, especially deep-sea wave power, were poor, and at best long term (ETSU 1992). See Table 4.1 for his conclusions on cost. However, since very little new research work has been done since the 1982 ACORD decision to cut funding, these conclusions are perhaps not surprising.

Be that as it may, clearly, wave energy was off the agenda for now, and given privatisation, there were much larger issues to face. It would involve major structural changes, with many implications, including for

Table 4.1 Review of wave energy costs—the Thorpe report R72

Technology	Discount rate	
	8%	15%
ART Osprey (OWC)	7 p/kWh	10 p/kWh
Shore-line OWC	6 p	9 p
Seagoing OWC (NEL)	17 p	29 p
Duck	16 p	26 p
SEA Clam	8 p	12 p
Bristol Cylinder	12 p	20 p
PS Frog	6–9 p	9–14 p

ETSU (1992)

renewables (Elliott 1990, 1992). Certainly, from 1992 onwards, a new phase seems to have emerged for renewables, with the emphasis on commercial take-up of near-market options.

4.2 Privatisation and the NFFO

The roll-out of the privatisation programme had not been without its problems. In 1989, during the initial stages of the privatisation process, it had become clear that the nuclear plants, both new and old, would need to be subsidised, at least initially, if they were to survive in an open market. Despite the inevitable distortion of the market involved, subsidies and market protections were therefore arranged (Roberts et al. 1991).

The idea of providing a subsidy for nuclear power was seen by some as something of an anomaly. Nuclear power was hardly a 'new' technology needing 'market enablement' support. Should not the same type of support also be available to the genuinely 'new' energy technologies, such as those based on renewable sources?

In the event, a step was taken in this direction with the inclusion of renewable energy projects, along with nuclear power, in the '**Non-Fossil Fuel Obligation**' (NFFO). The NFFO, introduced in 1990, was a statutory obligation placed on the newly privatised English and Welsh Regional Electricity Companies. Similar schemes were introduced in Scotland (the Scottish Renewable Order) and Northern Ireland (the Northern Ireland NFFO).

Under the terms of the NFFO, the privatised Regional Electricity Companies had to buy in a specified quantity of non-fossil fuel generated electricity. In effect, a protected market was established for the output from the nuclear power plants and for some renewable energy projects. The extra cost, over and above the so-called pool price, basically the cost that would have been incurred by conventional fossil fuel generation, was met by a levy on the electricity produced from fossil sources. This surcharge, which initially was around 10–11% each year, was then passed on to electricity consumers.

In its first year (1990–1991), the '**fossil fuel levy**', as it was called, raised some £1.3 billion from consumers, 98% of it going to nuclear power. So renewable energy remained very much the poor relation. Indeed, some critics saw the renewables part of the NFFO as a 'fig leaf', distracting attention from the nuclear subsidy. It was especially provocative since, even with this subsidy, a sell-off was proving hard, and the nuclear plants were left in the public sector meanwhile. The nuclear part of the NFFO was initially set at 8553 MW, that is, the entire available nuclear capacity was contracted for, while a series of competitive project selection rounds were established for the renewables, with an overall target of 1500 MW of Declared Net Capacity (DNC) being set for the year 2000. (The DNC rating takes into account the intermittent availability of some renewable energy sources.)

However, the nuclear part of NFFO scheme met with some opposition from the European Commission, which indicated that it was concerned that the fossil fuel levy was basically just a subsidy for nuclear power, which contravened the EC's fair competition 'state aid' rules. In the end, a compromise was reached, with the nuclear part of the levy being set to run only up to 1998, at which point it would be withdrawn. Initially, this limit was also applied to the renewable part of the NFFO, but from NFFO-3 onwards (after the delayed privatisation of the bulk of the nuclear fleet), this constraint was removed.

The initial 1998 time-constraint did lead to some problems: developers in the early NFFO rounds had to bid at prices that would suffice given a short time-frame, and one that was shorter for Round 2 than Round 1. So the winning wind farm projects got 6 p/kWh in the first round in 1990 and then 11 pence in the 1991 round. This made it look as if renewables were getting more expensive. But, with the 1998 limit removed, in Round 3 and beyond, that problem was avoided and the total amount available also rose, so more projects could go ahead at lower prices. Contracts under the subsequent renewable NFFO schemes (i.e. from NFFO-3 onwards) were also set to run for 15 years, during which time the project would receive support from the levy. In all, there were five auction rounds.

As Table 4.2 illustrates, the nuclear part of the levy stayed more or less constant, but following the privatisation of the bulk of the nuclear plants in 1996, the allocations were cut back and then halted. By contrast, the small renewable component of the NFFO levy gradually increased, to a

Table 4.2 The fossil fuel levy 1990–95 £ million (rounded up)

	Total raised	To nuclear electric	For renewables
1990–1991	1175	1135	6
1991–1992	1324	1281	12
1992–1993	1348	1291	29
1993–1994	1233	1139	68
1994–1995	1204	1081	96

Source: SERA (1995)

Subsequent renewable NFFO figures:

1995–1996	95
1996–1997	113
1997–1998	127
1998–1999	127
1999–2000	56
2000–2001	65
2001–2002	54

Source: Wilson (2001)

peak of £127 million. But with the initially large nuclear part winding down, the *overall* levy was reduced: by 1997, the levy surcharge was running at around 2.2% instead of 10–11%, and it continued to run, though with no more auction rounds, at reduced levels, up to 2002, when the scheme was closed and (as we shall see) replaced by a new system.

The NFFO scheme was seen only as an interim measure deigned to support 'near market' projects. Once the technology involved had matured, subsequent projects should be able to succeed commercially without levy support. The government was therefore looking for 'cost convergence' with conventional fuels, that is, reduction in costs so as to reduce the need for NFFO support in future.

The main 'near market' renewable energy-based technologies to benefit from the NFFO were wind power, small-scale hydro and various energy from waste technologies, notably landfill gas and waste combustion. By mid-1996, contracts had been placed under successive rounds of the NFFO for around 320 MW (DNC) of wind capacity, 554 MW of waste combustion projects and 169 MW of landfill gas projects, together with 56 MW of small hydro. NFFO-4, announced in February 1997, added a further 340 MW DNC of wind contracts, 241 MW of waste combustion contracts and 174 MW of landfill gas contracts, plus 13 MW of small hydro.

Cost convergence seemed to be well in hand. For example, as noted above, wind projects were given a premium price of 11 p/kWh in the second round of the NFFO (in 1991), but by the time of the third round (in 1995), they were only given an average price of around 4.35 p/kWh, with some projects in Scotland operating at 3.79 p/kWh. In the fourth NFFO, in 1997, wind projects received an average price of 3.53 p/kWh, with the lowest contracted price being 3.11 p/kWh. Landfill gas projects were given an average price of 3.01 p/kWh in NFFO-4, with some schemes being offered 2.79 p, while some waste combustion projects were given 2.66 p/kWh. The overall average was 3.46 p/kWh. Prices were to fall even further in the last, very large, 1177 MW NFFO-5, in 1998, to an overall average of 2.71 p/kWh. For comparison, the average pool price (i.e. for a conventional electricity) for the 12 months to 31 March 1998 was 2.67 p/kWh (Renew 1998a).

However, it has to be said that these NFFO prices were all *contracted* prices and not all the projects that won contacts managed to deliver. There were some drop outs. Some developers had evidently bid too low and were unable to proceed in practice. This seems to be an inevitable risk with competitive price/capacity auction type systems. Even if penalties for non-delivery are included, investors may still speculate. The result with the NFFO was that less capacity was actually delivered than had been agreed (Mitchell 2000). Nevertheless, the competitive element did mean that, overall, costs were pushed down. Moreover, the scheme did offer some renewables a way forward, even if that had initially meant riding on the coat-tails of nuclear.

4.3 The Falling R&D Budget

While the NFFO provided ongoing support for near-market renewables, funded by consumers, projects needing more R&D support were not so fortunate. Moving back in the story to 1994, by then renewables had received a cumulative total of £232 million in R&D support, having reached a peak of £25.6 million in 1992–1993 (see Table 4.3). However, as noted earlier, for the next phase, Energy Paper 55 had indicated that the R&D budget would be progressively reduced.

Table 4.3 Departmental expenditure on renewable energy 1975–1997
Expenditure in **£ millions** (in money of the day)
This updates the data in Table 3.1, which included estimates

1975–1976	0.5
1976–1977	1.0
1977–1978	2.5
1978–1979	3.6
1979–1980	8.2
1980–1991	11.3
1981–1982	17.3
1982–1983	14.7
1983–1984	11.7
1984–1985	14.9
1985–1986	13.2
1986–1987	12.0
1987–1988	16.3
1988–1989	16.0
1989–1990	17.9
1990–1991	20.8
1991–1992	24.8
1992–1993	25.6
1993–1994	25.2
1994–1995	19.8
1995–1996	17.8
1996–1997	13.3 (estimated)

Source: DTI data (Hansard 1988, 1996)

This policy was confirmed in March 1994, with the publication by the DTI of Energy Paper 62, *New and Renewable Energy: Future Prospects in the UK*, which announced a 20% cut for public R&D on renewables (down from £25.6 million as in 1992–1993 to £19.78 million for 1994–1995), with further cuts expected in subsequent years (DTI 1994a). In the event, this is what has happened: the 1995–1996 allocation was £17.8 million and the 1996–1997 allocation £13.3 million. See Table 4.3.

Figure 4.2 shows the DTI's expectations for the Research, Development, Demonstration and Dissemination (RDD&D) expenditure in future, with the private sector expected to play its part, along with the EC, but with overall funding falling, much as had been projected in EP55 (Fig. 4.1). So it was a major cut-back and refocusing decision, with the post-privatisation decks being cleared and the DTI's commitment being

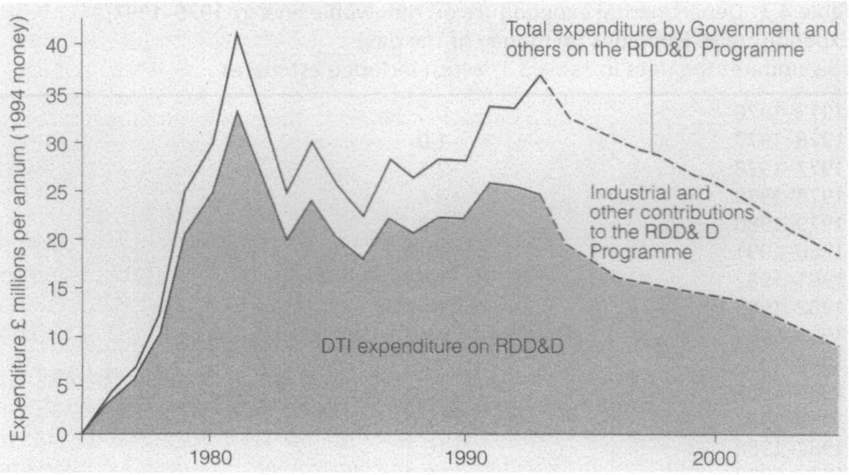

Fig. 4.2 Funding pattern for RDD&D for New and Renewable Energy technologies. (DTI 1994a)

reduced. EP 62 confirmed the wave energy close down, but, surprisingly, included a proposal to wind up the deep geothermal and tidal work, although, on the latter, the door was left open to developers to continue unaided.

As Energy Minister Tim Eggar put it, 'Resources will be concentrated on key technologies with good prospects of commercial application in the coming decades – such as solar, energy from coppice and waste, wind and fuel cells. Existing work on Passive Solar Design will be completed over the next few years and DTI involvement then reduced as the technology matures. The Government's geothermal R & D and tidal programmes will be closed when current commitments end. The existing work on wave energy will be completed, but no further commitments undertaken, as this technology has limited potential to contribute commercially to energy supplies in the next few decades' (Eggar 1994).

Putting this in context, Box 4.3 presents the DTI's classification of the technological options from EP 62. The first category contains technologies that were seen as 'near market' and which could be supported by the NFFO, along with some RDD&D. The second category covers technologies which would be further assessed and possibly supported with

RDD&D, including photovoltaics, fuel cells and photoconversion (advanced techniques for the biological or chemical conversion of light into useful energy). The third category contains the 'long shots' (e.g. wave, tidal, offshore wind) which would be left, at best, to the long term, on a 'watching brief' basis.

Box 4.3 DTI Classification of Renewables in EP 62

Market Enablement via NFFO and/or RDD&D
Passive solar design
Agricultural and forestry wastes
Municipal and industrial wastes
Advanced conversion technologies
Hydro power (small, new)
Landfill gas
Wind power (onshore)
Energy Crops
Assessment, RDDD&D
Photovoltaics
Photoconversion
Advanced fuel cells
Active solar
Watching Brief
Wave energy
Geothermal hot dry rock (HDR)
Geothermal aquifers
Wind power (offshore)
Tidal power
Hydro power (large, new)
and other technologies not listed.

Source: DTI (1994a)

For the near-market options, EP62 saw the NFFO as 'becoming the prime mechanism for support' and hoped that it would be possible to draw out £3 billion from the private sector over the next 10 years. But it accepted that there was also a need for a government-supported RDD&D programme, to 'assess and develop technologies, remove inappropriate market barriers, ensure the market is fully informed and encourage internationally competitive industries to develop'.

4.4 Interim RDD&D Appraisals: ETSU R 82/EP61

While Energy Paper 62 relayed the main policy conclusions, a wider overview was provided in a parallel ETSU report, R82, 'An Assessment of Renewable Energy for the UK' (ETSU 1994), which laid out the basis for the resource and cost estimates in EP62 and provided a justification for the policy decisions made in EP62. R82 was in fact the most comprehensive report on renewables in the UK that had been produced so far, in effect summarising the results of the UK renewable R&D programme to date. It set its technology review within the context of a range of supply and demand scenarios, and looked at the opportunities and needs for further RDD&D for each option, with a series of 'Technology Modules' providing the detail. These Modules are discussed in Box 4.4.

The scenarios ranged from optimistic to pessimistic (e.g. in terms of oil prices, carbon taxes and so on) and included a nicely named mixed scenario, 'shifting sands'. On the best scenario for renewables, renewables would supply 60% of UK electricity by 2025; on the worst, the contribution would be very small. Its Appendix 2 brought this data together in a standard RDD&D appraisal format, compatible with the wider appraisal of all UK energy options reported in Energy Paper 61, which, confusingly, came out after EP62, and which is reviewed below.

The renewables appraisal in R82 first identifies full RDD&D requirements for each contribution, and then the opportunities for new DTI-supported RDD&D effort are assessed. It makes for bleak reading in places. Thus tidal power had 'limited prospects in the UK', so there were no further development, demonstration or dissemination requirements, although some limited further research, for example, on environmental impacts, could be carried out. There was however a small 'let-out' clause, suggesting that it might be worth finding out 'whether reverse economies of scale apply to tidal barrage projects', so as to 'see if small schemes are more financially attractive'. But given the assumption of 'zero commercial potential by current UK criteria', overall, no real RDD&D opportunities were identified. So tidal power, once a front runner, was now seen as a 'watching brief' long shot, with, in any case, given privatisation, large barrages being seen by the DTI as unlikely to be taken up by the private sector, while the government would not provide support for them. Although, as we shall see, some attempts to get support were nevertheless made later, it seemed clear that there was little enthusiasm left from the DTI.

Following on from the results of the Thorpe report (ETSU R72), which had been published in 1992, R82 also concluded that wave energy, another one-time leader, had 'limited prospects', essentially just for shoreline systems, with a small RDD&D programme being envisaged. This idea however seems to have been dropped in the EP62 policy report, which, as we have seen, cut all R&D on wave power.

Geothermal Hot Dry Rock technology, another one-time leader, was now seen as having 'poor prospects', with just a 'watching brief' being proposed, as part of the EC collaborative R&D programme, but with no demonstration projects and only minimal information dissemination being necessary. It had clearly fallen from grace, presumably due to the poor results from the Cornish project. Overall, the assumption was, as with wave power, of 'zero commercial potential', with no significant UK RDD&D opportunities being identified. Geothermal aquifers, support for which had been cut in 1985, were also in effect written off finally. Support for the Southampton aquifer project had already been halted, although the local council continued work on it, as part of a district heating system.

A little more positively, a limited photovoltaics support programme was seen as sensible, although the RDD&D requirements were claimed to be already 'largely addressed in DTI and/or complimentary industrial programmes', leaving only a need for some further basic work on new semiconductor materials, and development work to reduce cell costs and improve system operation, with niche markets like remote off-grid locations in mind, these possibly having export potential.

Wind power was similarly seen as having few requirements for further RDD&D effort, with much of it already being assessed as covered by existing DTI/industrial programmes, leaving just resource studies, system integration studies and advanced turbine development as opportunities not so far fully covered. A large potential resource was identified, along with a significant export potential, but little extra effort seemed likely to be forthcoming, with the huge offshore resource still being effectively ignored.

The 'largely addressed by existing DTI/Industry programmes' formula was repeated regularly, for example, for municipal and agricultural waste, hydro and passive solar, all of which were seen as having similar potentials. Passive solar, a one-time leader, was now seen by the DTI as no longer needing support. Similarly, for active solar, landfill gas and for energy crops and advanced biofuel conversion techniques, although these latter two were seen as having much larger potentials. Only photoconversion was seen as sufficiently novel to need significant new RDD&D support.

ETSU said that, in principle, the appraisals of opportunities for RDD&D were meant to look at options 'where risks may be too high' and 'where the potential returns are too long term' for conventional commercial support to be forthcoming. Overall, it concluded that 'most renewables are not currently competitive under current accounting systems' with large-scale introduction awaiting 'the anticipated reduction in renewable energy generating costs and the adoption of accounting systems that address the environmental costs of energy generation', for example, a carbon tax or something similar.

This was perhaps a fair statement of the market situation, but it was also a little dispiriting since a carbon tax looked some way off, and, arguably, the cost reductions would require more RD&D, which is just what, in the main, EP62 avoided, no doubt with privatisation in mind. Nevertheless, R82 was something of a landmark report, with a vast amount of detail in its 'Technology Modules', reviewing the state of play for each technology. Some of the highlights are given in Box 4.4.

Box 4.4 The R 82 Technology Modules (ETSU 1994)

Wind power

Wind was seen as a major resource, with the land-based theoretically accessible resource put at 120 TWh/yr for England and Wales, 190 TWh/yr for Scotland and 33 TWh/yr for Northern Ireland. The offshore accessible resource was put at around 380 TWh/yr, greater than then current total UK electricity usage (300 TWh/yr). However, the *maximum practical resources* were less, 32 TWh/yr on-land, given system integration problems, and 140 TWh/yr for offshore, although the cost of generation offshore was only put at 30% extra. Even so, R82 claimed, overall, that 'nowhere is the Technology cost-competitive with conventional power technology', although it did note that costs were falling.

Tidal

No economies of scale were identified between large and small barrages, but the 64 MW Wyre barrage site did have a slight advantage over the 8.6 GW Severn Barrage, with estimated generation cost, at 8% discount rate, of 6.8 p/kwh compared to 7.2 p/kwh for the Severn. The Mersey (700 MW) came out at 7.1 p/kWh, Conwy (33 MW) at 8.7 p/kWh, while results on Duddon (100 MW) had yet to emerge. R82 noted that, although so far the Wyre looked the cheapest, 'Ministers have declined to support further development of the project'. Only passing mention was made of the tidal stream option, which was seen as likely to be very expensive, as an earlier ETSE report had suggested (ETSU 1993a, b).

Wave

In line with the 1992 Thorpe review (R72), offshore deep-sea wave was discounted as expensive and long term (10 TWh/yr by 2025 at best), and the 'shore-line' resource was seen as small (1.7 TWh/yr accessible resource). The new 1 MW near-shore Osprey device, then being supported by the European Commission, was not mentioned specifically, but the EC's programme was noted.

Photovoltaics

Although the existing PV contribution was 'relatively small', R82 noted an assessment that the theoretical potential for integrated roof systems, using all accessible surfaces and orientations, could be up to 200 TWh/yr, and as much as 360 TWh/yr by 2020. More than total UK power use! But it said costs were still high, and, for south facing units, it put long-term resource, at a cost of less than 10 p/kWh, at only 84 TWh/yr, at 8% discount rate, or just 4.2 TWh at 15%.

Energy Crops

Arable coppicing, although in its infancy, was favoured in the medium term, following on from the existing NFFO-supported waste conversion projects, with Gasification of crops and agricultural wastes now being seen as 'ready for demonstration in the UK'. Indeed, R82 identified a range of new options in the biofuel field, which it saw as a key new area for development, although it did recognise that there might be some environmental objections.

Given the diversity of energy crops, costs were often difficult to generalise, and obviously depended on location and conversion technology, but 3–4 p/kWh was seen as typical, with the maximum practical resource put at up to 200 TWh/yr by 2025, plus smaller amounts from wastes.

Of the rest, Geothermal **Hot Dry Rock** was seen as an 'immature' technology, with costs put at 17 p/kWh at 8% discount rate, **hydro** was seen as small but relatively cheap, with more than 300 MW of new small-scale hydro capacity available at less than 2.7 p/kWh, assuming a 10% rate of return. On the heat side, **passive solar** remained a popular option, **active solar** less so, but the big resource for heat as well as power was seen as being in the biofuel/waste area.

In May 1994, the DTI published the results of a comprehensive review of *all* the energy supply technologies, *Energy Technologies for the UK*, Energy Paper 61. It drew together all the separate studies, including those on renewables, and looked at future prospects to 2025 and at R&D options (or rather RDD&D—Research, Development, Demonstration and Dissemination), all based on a range of energy supply/demand scenarios, including a 'heightened environmental concern' scenario (DTI 1994b). It classified the technologies under four main headings, robust, vulnerable, fragile and unpromising, depending on how well they did in the various scenarios to 2025.

Unsurprisingly, given the previous reviews of renewables (EP 62/ R82), tidal barrages, offshore wave, offshore wind, geothermal aquifers and HDR all came under the last category, that is, **unpromising**. Hydrogen production by electrolysis, hydrogen use in internal combustion engines and fuel cells for large-scale generation were also put in this category, which also, interestingly, contained nuclear fusion. All of which offered 'no contribution on any scenario/discount rate', except perhaps under an environmentally constrained future.

Robust technologies were seen as offering a potential contribution under all scenarios, and at all discount rates: of the renewables, only passive solar came in this category. The **vulnerable** technologies included wind power and municipal solid waste in the short term, hydrogen from biomass and photovoltaics in the medium term, these possibly offering a potential in three or more scenarios, regardless of discount rate. The **fragile** technologies, which offered contributions under only one scenario, regardless of discount rate, included landfill gas, agricultural waste as fuel and shore-line wave power in the short term, energy crops in the medium term and biodiesel in the long term.

EP61 then looked at the implications for RDD&D on the basis of a 'risk versus payback' assessment. It reported that the bulk (55%) of the 1992–1993 RDD&D expenditure (both public and private) was in the high-risk/low-payback zone and commented that 'plans have already been announced to curtail some of these programmes', quoting tidal power and geothermal HDR as examples and also mentioning the closure of the fast breeder nuclear reactor programme. These, EP 61 noted, accounted for 75% of the high-risk/low-payback expenditure in 1992–1993. It also included nuclear fusion in the high-risk/low-payback list, but said, a little mysteriously, that 'fusion has long been acknowledged to have objectives other than cost-effective energy supply'. It certainly seems to have had a charmed life, attracting a large part of the energy R&D budget over the years.

4.5 Looking to the Future: Technology Foresight

In the various reports mentioned above, ETSU had provided what might seem to be an exhaustive analysis of the energy supply options, but there still remained some uncertainty as to which technologies were the best bet for the future. Given the difficulties already experienced in trying to 'pick winners' and deal with risks at the R&D stage, the government also supported what is sometimes seen as an alternative approach to technology assessment, the use of Technology Foresight (TF) analysis. In 1993, the UK Governments Office of Science and Technology (OST), a policy unit at that time based within the Cabinet Office, announced that it was going to introduce a 'Technology Foresight' programme, which would look 20 years ahead in selected fields, so as to provide guidance on future R&D priorities (Elliott 1996).

In early 1994, 15 Sector Panels were set up, overseen by the Office of Science and Technology, bringing together a wide range of expertise from the civil service, industry and academia. Each panel carried out a Delphi 'opinion refining' exercise. For example, the energy panel was invited to comment on the prospects of the price of photovoltaic solar cell modules falling below $2 per peak watt, in the 20-year TF time-frame. Panel members were then asked to revise their original estimates and projections in the light of those of other participants, so that a consensus could be reached.

The results from the UK TF process emerged in April 1995: 15 reports were published covering each sector, together with an overview by the OST. The *Energy* sector study noted that protecting the environment was 'the single most important issue' shaping the development of energy technology, and it identified nine specific area of development. The majority involved conventional fossil fuel technology: for example, better oil recovery and gas combustion techniques were called for. However, renewable energy technology was also considered quite relevant. In direct contrast to many of the earlier studies, the *Energy* TF gave solar photovoltaics a high priority (OST 1995).

While PV was strongly placed by the *Energy panel*, there were some differences in treatment and emphasis amongst the various sector studies. Thus, the *Agriculture, Natural Resources and Environment* sector study prioritised wind, wave and tidal power, whereas wave and tidal power technologies, along with the use of hydrogen as fuel, geothermal energy and ocean thermal systems, were dismissed by the *Energy* sector study as unlikely to be relevant within the 20-year TF time-frame. In effect, the *Energy* study thus followed the lines adopted by the DTI in Energy Paper 62.

Similarly, there were disagreements about energy crops. The *Energy* study put them in their second ranking, while the *Agriculture, Natural Resources and Environment* study gave them top priority. The latter study concluded that the UK was weak at developing energy-efficient housing, an area which was also seen as having high priority in the *Energy* study, while the *Construction* sector study was more sanguine, although it did feel that the UK's construction industry was technically weaker than many of its international rivals (Elliott 1999a).

Summarising and synthesising the sector reports, the TF steering group's overview report suggested 11 'key topic areas' for the UK, these areas of technology being seen as 'so promising that further work was absolutely vital and demanded top priority'. They included Environmentally

Sustainable Technology, including 'alternative, sustainable energy technology and resource conservation' (OST 1995).

Given that broad framework, and the more detailed TF outputs, subsequently, solar photovoltaics received some support under the Foresight Challenge scheme. With strong lobbying from pressure groups like Greenpeace to press ahead with PV, and commercial interest growing worldwide, it did seem that PV could be the next major renewable to breakthrough, although, as we will see, that took some while.

The other major possible breakthrough option could be energy crops. As noted above, that had got some support from the TF exercise, although so far, the favoured option, short rotation coppice (SRC), had been relatively slow to progress: its prospects seemed linked with the outcome of the debate over the EU Common Agricultural Policy. In the event, as we shall see, little has happened in the UK, the energy crops area.

Finally, there were the various offshore options, which you would think would be very relevant to a maritime nation like the UK. Offshore wind looked like it would lift off at last, given that one large project had obtained support under NFFO-4, but wave power and tidal power seemed stalled. As we shall see, offshore wind did lift off significantly, while marine renewables (wave and tidal stream) still languished.

However, as may perhaps by now have become a familiar pattern in the story, there was then a rethink. One wave energy project, Wavegen's 75 kW shore-line mounted Oscillating Water Column device, had done quite well, with EU funding being provided for a 300 kW follow-up, the so-called Limpet. With progress like that as a background, a more positive assessment of wave energy began to emerge, with, in 1998, a new study by Tom Thorpe, who had been ETSU's wave energy project officer (you may recall he was the author of their 1992 R72 wave energy report), and was now with AEA Technology, a spin-off company that had emerged from the UKAEA. He noted that there has been 'a considerable improvement in the costs of devices, so that there are now several with costs of 5p/kWh (or less) at 8% discount rate' and, added, 'this indicates that wave energy is already economically competitive in niche markets and has a good prospects of being commercially competitive with further R&D' (Thorpe 1998).

This fed into the TF process, including to a one-day workshop on Future Opportunities for Marine Renewable Energy, organised by the newly convened Foresight Marine Panel in May 1998. It was chaired, perhaps symbolically, by Gordon Senior, an independent consultant, who you may recall had earlier fallen foul of the government, when he claimed that the conclusions to a report that he wrote supporting wave power had been tampered with (Renew 1998b).

In its subsequent final report, the Marine Foresight panel was quite positive on wave power, and also, though less so, on tidal stream technology (OST 1999). In response, Labour's energy Minister, John Battle, announced a new wave energy programme, as part of a £43 million renewables R&D programme over the next three years, as proposed in a new consultation paper, 'New and Renewable Energy: prospects for the 21st Century' (DTI 1999). So it seems the next century was going to be much like the last one, at least in terms of the cycles of ups and downs faced by wave energy.

4.6 The End of the 1990s: Liberalising Markets

Even with the help of TF analysis, it had proved hard to get agreement on which technologies to support and ever harder to agree on how to fund them. The range and intensity of the debates had been striking, particularly in the early years, when individual ministers and officials were much to the fore, and Select Committees were regularly challenging them. Although somewhat repetitive battles over technology choice persisted, in the main, after privatisation and the demise of the DEn and the Select Committee on Energy, there was a change. The emphasis shifted from overall policy statements and wider debates, to the presentation of details of the latest NFFO and relatively minor squabbles about them.

Certainly, there were problems with the NFFO, and it was important that these were addressed. But the focus was on relatively short-term market concerns. Although it is clearly important to devise mechanisms to create and support a market for existing green energy systems, arguably there is also a need to look to the longer term and develop technologies for the next stage. However, as we have seen, the basic approach adopted by the government had involved relying on markets to decide on which options to follow. That view came strongly to the fore again in the final phase of the Conservative Government's proposed electricity market reform, planned to start running from 1998, involving the liberalisation of the power market. In the event though, it did not get a chance to see it through, given that it lost power to Labour in 1997.

Perhaps surprisingly, the liberalisation plan was continued under Tony Blair's New Labour Government. It was evidently seen by New Labour as part of its economic modernisation programme. The Blair administration also set about replacing the NFFO support system with a new, arguably even more market-orientated system. It involved the introduction of tradable Renewable Obligation Certificates (ROCs), awarded to companies who met the renewable targets set for them. That created a new market.

We will be looking at the way the ROC system panned out, as it got underway in 2002, in the next part of this book, but, by providing new market incentives, it was meant to fit into the overall market liberalisation approach, with new players incentivised to enter the market. However, it was unclear whether reliance on market liberalisation would necessarily ensure diversity, or security of supply, much less the expansion of renewables. Arguably, that would require other measures (Elliott 1994a).

Nevertheless, liberalisation might benefit consumers. It was claimed that it should make it possible for consumers to purchase power from any provider, with the national power grid acting as a common carrier. There was certainly a flurry of 'green energy' tariff schemes from the power utilities, who offered to match any power used by consumers with renewable power they contracted to produce or buy in, usually in return for a small extra charge and subject to source verification rules (Elliott 1999b). In principle, this new green market could enable a whole range of small companies to enter the field, increasing competition and forcing prices down. A few small companies did try their luck, and some of them (notably Good Energy and the Renewable Energy Company, now Ecotricity) went on to become successful green power suppliers and retailers, with Ecotricity in particular able to support some new capacity. Nevertheless, as far as the large power companies were concerned, the new green tariff retail market was marginal. They engaged in it mainly to avoid losing customers to rival schemes (Elliott 2000).

As we shall see, in the 2000s and beyond, although some diversification initially took place, the main result of market liberalisation has been market concentration, with a few large companies dominating. Renewables nevertheless managed to find ways through, not least since they were becoming increasing economically attractive. You might say markets have worked in the end, although, as we shall see, they can also slow and divert progress.

So of course can public opposition, and this has arguably been enhanced by market pressures. In the more competitive market created by privatisation and liberalisation, there were inevitably pressures on developers to site wind farms in more profitable high wind speed locations, most of which were more visually intrusive. So, with the number of suitable sites also diminishing as the programme expanded, there was an increased potential for local planning conflicts.

Some of the wind farms were certainly opposed by local people, and although the scale of this opposition can be debated, it did lead to some negative headlines and articles, and to a slow-down. One columnist even claimed that 'conventional rural opinion is now probably, if anything, anti-wind power' (Ridley 1994). The reality was somewhat different. While in some cases there was a noisy minority opposed, it was rarely a majority. See Box 4.5 for some examples. Even so, as we shall see, the view that wind farms were unpopular played a role in the next phase of the renewable programme, particularly when, later, a Conservative Government was elected.

Box 4.5 Results of Some Public Opinion Surveys on Wind Farms

Kirby Moor

A study of the views of 250 local residents near the 12 turbine wind farm at Kirkby Moor in Yorkshire was commissioned in February 1994, 6 months after start-up by National Wind Power. It found that:

- 82% supported the development of wind farms in the area and 84% thought that more energy should be generated from renewable sources;
- 83% were 'not at all concerned' or 'not very concerned' about the noise that they make;

- Of those who could see the wind farm from their houses 77% were 'not at all concerned' or 'not very concerned' about the impact on the landscape.

Taff Ely

A study of views from 250 local residents near the 20 turbine wind farm at Taff Ely in Wales was commissioned in February 1994, 6 months after start-up, by East Midlands Electricity. It found that:

- Only 2% strongly opposed the development of wind farms in the areas;
- 75% said that either they could not think of any disadvantages of wind power or there were no disadvantages;
- Noise was not perceived as a problem with only 3% saying they could hear the wind farms from their homes.

Source: Data relayed by the British Wind Energy Association 1994, as reproduced in Elliott 1997 (and in the second edition, 2003) which also includes coverage of the responses by local planning bodies, the media and pressure groups, based in part on Elliott 1994 (Elliott 1994b, 1997, 2003)

However, that negative interpretation has to be set against the wider reality that renewables were mostly very popular. As noted earlier, in a Gallup poll in 1991, carried out on behalf of Friends of the Earth, 87% said they wanted the government to do more to increase the proportion of electricity coming from renewable energy and, as we shall see, support for onshore wind has been high in the more recent polls carried out by the government, running at 74–76%.

That is not to say that local objections to renewable energy projects can or should be ignored or down-played. Renewables were, after all, meant to avoid environmental problems, and wind farm developers were keen to do what they could to minimise any impacts. Most of them carried out extensive local consultation before going ahead, and post-construction surveys were also carried out, as illustrated in Box 4.5. One of the interesting results from some of these surveys was that, in cases where opposition had been relatively high when the project was first proposed, after it had been built and running for a while, opposition declined. In some cases, it may have been the fear of the unknown that was behind the initial opposition: actual experience can breed acceptance (Lee et al. 1989;

ETSU 1993b). Nevertheless, as we shall see, in the 2000s, public opposition, along with other factors, slowed the expansion of onshore wind in the UK. Clearly, the UK renewable story, although moving forward, was not going to be without continuing problems.

References

Department of Energy. (1988). *Renewable Energy in the UK: The Way Forward,* Energy Paper No. 55. London: Department of Energy.

DTI. (1994a). *New and Renewable Energy: Future Prospects in the UK,* Energy Paper 62. London: Department of Trade and Industry.

DTI. (1994b). *Energy Technologies for the UK,* Energy Paper 61. London: Department of Trade and Industry.

DTI. (1999). *New and Renewable Energy: Prospects for the 21st Century,* Consultation Paper. London: Department of Trade and Industry.

Eggar, T. (1994). *DTI Press Release 31/3/94.* London: Department of Trade and Industry.

Elliott, D. (1990). New Technology and Restructuring in the UK Energy Industry: The Case of Renewable Energy. *Technology Analysis & Strategic Management, 2*(3), 253–263.

Elliott, D. (1992). Renewables and the Privatisation of the UK ESI. *Energy Policy, 20*(3), 257–266.

Elliott, D. (1994a). UK Renewable Energy Strategy: The Need for Longer Term Support. *Energy Policy, 22*(12), 1067–1074.

Elliott, D. (1994b). Public Reactions to Windfarms: The Dynamics of Opinion Formation. *Energy & Environment, 5*(4), 343–362.

Elliott, D. (1996). Technology Foresight: An Interim Review of the UK Exercise. *Technology Analysis & Strategic Management, 8*(2), 191–199.

Elliott, D. (1997 & 2003). *Energy, Society and Environment.* London: Routledge, 1997 Edition On line at: https://docslide.us/documents/energy-society-and-environment-routledge-introductions-to-environment.html

Elliott, D. (1999a). *Technology Foresight and Renewable Energy: The Next Generation,* Technology Policy Group Occasional Paper 28. Milton Keynes: The Open University.

Elliott, D. (1999b). Prospects for Renewable Energy and Green Energy Markets in the UK. *Renewable Energy, 6,* 1268–1271.

Elliott, D. (2000). Green Power and the Liberalisation of the UK Electricity Market. *International Journal of Ambient Energy, 20*(1), 3–13.
ETSU. (1992). *A Review of Wave Energy*, The Thorpe Report, Energy Technology Support Unit, R72. London: Department of Energy.
ETSU. (1993a). *Tidal Stream Energy Review*, Energy Technology Support Unit, Report 05/00155. Harwell.
ETSU. (1993b). *Attitudes Towards Windpower: A Survey of Opinion in Cornwall and Devon*, ETSU, W/13/00 354/038/REP. Harwell
ETSU. (1994). *An Assessment of Renewable Energy for the UK*, R82, Energy Technology Support Unit. London: Department of Energy.
Hansard. (1988). *DTI Expenditure Data for 1975–88*. https://api.parliament.uk/historic-hansard/written-answers/1988/mar/11/renewable-sources#S6CV0129P0_19880311_CWA_54
Hansard. (1996). *DTI Expenditure Update to 1996*. https://api.parliament.uk/historic-hansard/written-answers/1996/apr/02/environmental-technology-industries#column_148w
Lee, T., Wren, B., & Hickman, M. (1989). *Public Responses to the Siting and Operation of Wind Turbines*. Robens Institute and Department of Psychology, University of Surrey.
Mitchell, C. (2000). The England and Wales Non-Fossil Fuel Obligation: History and Lessons. *Annual Review of Energy and the Environment, 25*, 285–312.
New Scientist. (1990, April 14). Sorry, Ducks. *New Scientist*. https://www.newscientist.com/article/mg12617122-300-sorry-ducks/
OST. (1995). *Technology Foresight: Progress Through Partnership*, Report No. 13. London: Energy, Office of Science and Technology, Department for Trade and Industry.
OST. (1999). *Energies from the Sea- Towards 2020*. London: A Marine Foresight Panel Report, Office of Science and Technology, Department for Trade and Industry.
REAC. (1992). *Renewable Energy Advisory Group Report to the DTI on Renewable Energy*, Energy Paper 60. London: Department of Trade and Industry.
Renew. (1998a). NFFO-5: The Biggest Yet. *Renew* 116, Nov–Dec. Renew On-line 17. http://eeru.open.ac.uk/natta/rol17.htm
Renew. (1998b). Sea Energy Rules Again. *Renew* 114, July–Aug, Renew On-line 15. http://eeru.open.ac.uk/natta/rol15.htm
Ridley, M. (1994, February 6). Column on Wind Farms. *Daily Telegraph*.
Roberts, J., Elliott, D., & Houghton, T. (1991). *Privatising Electricity: The Politics of Power*. London: Belhaven.

Ross, D. (1990, November 10). *Technology: Europe Misled Over Wave Energy*. https://www.newscientist.com/article/mg12817423-500-technology-europe-misled-over-wave-energy/

Select Committee on Energy. (1992). *Renewable Energy*. Session 1991–92, Fourth Report, House of Commons Select Committee on Energy, London.

SERA. (1995). *Data Reproduced in SERA's Journal New Ground 44, Based on a Parliamentary Answer*. London: Socialist Resources and Environment Association.

Spicer, M. (1988, January 13). Parliamentary Answer. *Hansard*, *125*, cc 96-8W/296W. https://api.parliament.uk/historic-hansard/written-answers/1988/jan/13/alternative-sources-of-energy

Thorpe, T. (1998). *Overview of Wave Energy Technologies*. Harwell: AEA Technology.

Wilson, B. (2001, November 21). Parliamentary Answer, Support for Renewables. *Hansard*, Col 300-01W. Holding Answer, November 7 Col. 306W. https://api.parliament.uk/historic-hansard/written-answers/2001/nov/21/renewable-energy#S6CV0375P0_20011121_CWA_200

5

Interim Analysis *of the Story So Far*

5.1 Interim Technological Review

It is helpful to pause in our account of UK renewable energy policy developments and programmes at this point to take stock, before we move on to cover the next phase, from the 2000s onwards. Looking back over the UK renewable energy programme as it had unfolded up to the end of the century, it is clear that some technologies had prospered while others had not. Wind, waste combustion and some biomass projects had done well, but most of the more novel options had yet to achieve a breakthrough.

Deep-sea **wave power** was initially a front runner amongst the UK's renewables. However, as we have seen, following the ACORD review in 1982, and some adverse economic assessments, work on deep-sea wave power was halted. This decision was the subject of much controversy, aided by journalist David Ross's influential 1979 book *Energy from the Waves* (Ross 1979), which was revised in 1981, and rewritten and updated as *Power from the Waves* in 1995 (Ross 1995). There were certainly doubts about the reliability of the ACORD assessment and some of the data used. Godfrey Bevan, the civil servant Director of the renewables programme, told the Select Committee on Energy that 'there is

D. Elliott, *Renewable Energy in the UK*, Energy, Climate and the Environment,
https://doi.org/10.1007/978-3-030-04765-8_5

definitely scope for different judgments at the early stage of the development of a device', which the committee evidently took to be an admission that 'a mistake was made in 1982' (Select Committee on Energy 1992). However, as we also saw, a subsequent review by Thorpe, published in 1992, came to similar conclusions to those reached by ACORD. As pointed out earlier, given that very little new work on deep-sea wave power had been done since 1982, perhaps that was not surprising, but being more positive, the Marine Foresight report, in 1999, argued that 'since so little work has been completed in this area, the learning curve is still steep and valuable results should be obtained from relatively small further investment in R&D' (OST 1999).

As noted above, some work had continued on smaller-scale onshore/shore-line devices (like the Limpet), using the Wells Turbine developed at Queens University in Belfast, but, although likely to be useful in some situations, for example, in developing countries, the energy potential for such systems was much less than for fully offshore devices. Moreover, the wave energy field had taken something of a knock by the very visible loss of the EU-supported 1 MW Osprey Oscillating Water Column (shallow) offshore prototype after a storm off the North of Scotland in 1997, while it was being positioned and filled with sand ballast for final sea-bed mooring. Stephen Salter had always said that tank tests were better, since, although you could get useful real-world experience from sea tests, at this stage there were risks.

Even so, wave energy was not dead, with three projects going ahead under the Scottish version of the NFFO, and yet another ETSU report (R122), produced as part of the new DTI reviews of renewables (DTI 1999), putting the theoretical resource at 100–140 TWh p.a. for near-shore/onshore devices and 600–700 TWh p.a. for offshore systems (ETSU 1999). ETSU suggested that in practice, onshore and near-shore wave devices might supply up to 2.5 TWh p.a., while the practical offshore potential was put at 50 TWh p.a. Overall, ETSU saw wave power making a significant showing by 2025, with the cost-resource curves at 8% discount rate cutting the x-axis at 3 p/kWh. That did look a little more hopeful. As we have seen, the Office of Science and Technology's Marine Foresight panel also offered positive conclusions, and the then energy Minister, John Battle, indicated enthusiasm for wave energy. Nevertheless, wave energy was still quite marginalised, with it only being shifted from the 'very long term' category in the DTI's ranking, to the 'long term' category, and similarly for tidal stream, with for both options, the 1999 Marine Foresight report noting that there was a need to 'overcome the setback for this form of energy created in the 'public mind' by a perceived lack of success to date' and also to overcome 'prejudice due to the earlier poor performance in some cases' (OST 1999).

Following the 1982 ACORD review, **Wind power**, initially felt by some to be of marginal importance, became a front runner, and a £54 million R&D programme followed, with the emphasis on large turbines. As we saw, a 3 MW unit was built by the WEG on the Orkneys. As it turned out, large machines did not prove to be what the market required at that point. As the NFFO got underway, the emphasis shifted to machines of around 300 kW, although the scale then increased, with machines of 1 MW and above emerging, in what was to become a quite rapidly expanding programme. Offshore wind had been strongly placed in the OST Marine Foresight study, and, as we shall see, that too later began to lift off significantly, with even larger machines. The UK's first offshore wind farm, at Blyth in Northumberland, opened in 2000, with two 2 MW turbines.

As we have seen, the **Tidal** option was at one time assessed as quite significant. The UK had some of the best sites for tidal barrages in the world, notably the Severn Estuary. The overall UK tidal potential was put at around 20% of its electricity requirements. However, the capital costs would be very large, for example, around £10 billion for an 8.6 GW tidal barrage on the Severn. Estimates later rose to £20 billion or more. A project on this scale might have been viable as a long-term publicly financed national investment, but, following privatisation, it seemed clear than no money would be available. Attempts were made to investigate smaller, less expensive tidal barrage options, notably on the Mersey, and the Mersey Barrage Company tried to obtain NFFO support for their scheme. In the end, however, this was unsuccessful, and several other smaller barrage proposals were also abandoned. Following DTI Energy Paper 62, the tidal barrage programme was mostly wound up, based on what amounted to an obituary for tidal power from ETSU in 1994 (R82), although, as we shall see, that may have been premature.

Barrage enthusiasts were still to have one more try (in the 2010s), and interest remained in the exploitation of **tidal streams**. It had been noted that the UK could in principle obtain perhaps up to 19% of its electricity from this source, although an ETSU report, published in 1993, suggested that the costs would be prohibitive (ETSU 1993). As a result, apart from a small independently funded experimental prototype, no further work was funded in this area in the UK, until a revival of interest was sparked by the 1999 OST Marine Foresight study.

Whereas tidal stream systems were novel and untried, the primary problem in the case of tidal barrages was not technological: the technology existed and was relatively mature, as witness the existing 240 MW Barrage on the Rance estuary in France. The problem facing the UK barrages was finance, and the relatively short-term economic perspective that prevailed in the energy sector, as elsewhere, following privatisation.

By contrast, the problems that emerged with the **hot dry rock geothermal** programme were mainly technical. The potential resource was quite large, perhaps 10% of UK electricity requirement, although the technology was untried. It also turned out to be risky. The initial results from an experimental well at Camborne, in Cornwall, were disappointing: less power was produced than expected due to problems with the geology, and, following the DTI Energy Paper 62 1994 review, the UK programme was halted.

While the future for geothermal energy, at least in the UK, looked limited, the prospects for **Solar Photovoltaics (PV)** seemed to have improved. In one of its initial reviews of the renewable options (Energy Paper 16) published in 1976, the DEn considered solar cells as 'not relevant' to the UK, and this view prevailed for some years. However, during the 1980s, large PV research programmes were mounted in the USA, Germany and Japan, and in 1991, an ETSU report (R-50) suggested that the technology was worth reconsidering. A relatively small UK support programme followed, reaching around £1 million p.a. in 1994. Subsequently, as noted earlier, PV was backed in a TF assessment, and some further funding was provided, including £1 million from the Foresight Challenge fund for a PV implementation/demonstration programme in schools and colleges. But it was fairly minimal. It was not until much later that PV began a spectacular lift-off in the UK, using imported cells.

Finally, there were the various **biomass**-related energy technologies. Government funding had been provided for development work on the combustion of domestic and industrial **wastes**, and the collection of **landfill gas**, and, as already noted, several waste combustion and landfill gas projects subsequently obtained support in the successive rounds of the NFFO. The latter produced amongst the cheapest renewable power available.

In the third round of the NFFO, three **energy crop** projects also obtained support. These involved SRC plantations which supply wood chips which were then burnt to produce hot gases to power gas turbines for electricity generation. The SRC gasification projects received an average price of 8.65 p/kWh in NFFO-3, but by the time of NFFO-4, in 1997, 7 biomass gasification projects were supported with an average price of 5.51 p/kWh, so price convergence (on conventional power costs) was evidently underway. Certainly, it was felt that this technology could represent a potentially very significant new area for development. Indeed, at one point, in EP 62, it was estimated that, if its full potential could be developed, SRC might eventually supply *up to half* of the UK's electricity requirements (DTI 1994).

That turned out to be a little optimistic. The flagship £30 million project had been ARBRE, a pioneering 8 MW combined cycle wood chip/SRC gasification plant in Yorkshire, work on which started in 1998. However, it proved hard to get the complex new systems to work together effectively, and it was eventually abandoned in 2003, after only running briefly, leaving in its wake extensive, but now redundant, SRC plantations that had been started up ready to feed it (Boyle and Brown 2003). It was a major setback for SRC and gasification (Piterou et al. 2008). Subsequently, in the 2000s, some large-scale bio-conversion projects did go ahead, using large old coal plants converted to burning biomass (e.g. at the DRAX plant in Yorkshire). They have been seen as vital by the government given their large outputs, although they have mostly used imported forest-derived biomass, with, as we shall see, that attracting strong environmentalist objections.

By way of comparison and context, brief mention should perhaps be made at this point of the state of play with nuclear power. It had not benefitted from privatisation. In 1995, the Conservative Government had transferred the new Sizewell PWR plant, along with all the old AGR plants, to a new public company, British Energy, which was then privatised in 1996. Unfortunately, British Energy then ran into financial problems, worsened by the so-called dash for gas, and had to be bailed out by the government. It was eventually to be sold off to EDF.

With nuclear evidently facing economic problems, the prospects for new nuclear looked poor. In 1995, after a major review, the Conservative Government concluded that 'providing public sector funds now for the construction of new nuclear power stations could not be justified on the grounds of wider economic benefits and would not therefore be in the best interest of either electricity consumers or tax payers' (Nuclear Review 1995).

By contrast, despite the problems identified above, the prospects for some of the renewable technologies were looking reasonably good. Some had moved from the R&D stage to commercial uptake. It is worth looking briefly at some of the institutional background to this process, since it shaped how the technologies developed and the technological focuses that were adopted.

As we have seen, in organisational terms, by the end of the century, the UK renewable energy technology programme had become increasingly industry-led, with large companies playing key roles, although many independent consultants and academic experts were also involved, some of them sympathetic to, or even past members of, the earlier Alternative Technology movement. There was also still room for new small technology-led entrepreneurial start-up companies, sometimes spun out of University research groups. That was the case for some of the wave and tidal stream projects. Some new enterprises also spun off from the CAT, including Dulas Engineering, in 1990, and Ecogen, a wind farm developer, in 1999. In addition, some AT enthusiasts became active in local community energy groups, local support agencies and campaign groups. There was certainly a lot going on at the grass roots, even if funding was limited (NATTA 2000).

Given the funding issues, much of the technological focus for the local agency advice and support groups was on usually lower-cost and easier-to-initiate energy-saving projects, aided by the government-supported Energy Saving Trust (EST). However, its funding became increasingly limited, evidently (and tragically) a common fate for energy-saving programmes: see Box 5.1. Nevertheless, the EST did soldier on with energy saving and also supported some renewable energy projects (e.g. solar), providing advice on deployment, and in general acting as a base for access to support for local-level projects.

Box 5.1 Energy Saving Trust

The Energy Saving Trust was established by the Conservative Government in 1992 and was seen as a major part of their response to the commitment made at the 1992 UN Earth Summit to stabilising CO2 emissions by the year 2000. A range of energy-saving measures was to be supported, with a levy on consumer bills being used to fund them. The cost was expected to be substantial, £150–200 million p.a., totalling perhaps £1.5 billion by 2000.

However, it was not to be. The market regulator, which had regulatory discretion in the post-privatisation market, evidently did not think that a consumer levy was appropriate, and it was halted. In 1995, the EST programme was drastically cut back, with just £29 million provided for a limited programme (Owen 1995). The EST has nevertheless continued to operate, for example, providing a base for various subsequent government-funded consumer- and community-orientated advice and support programmes, very valuable, but at relatively low level. This episode might be seen as just one example of how energy saving always seems to be marginalised. We will be looking at some subsequent initiatives in Sect. 7.4. They mostly did not fare any better.

The scale of the community-based projects may have been relatively small, but, as well as providing practical help, these real-world local technology demonstrations, coupled with the wider outreach and campaigning work, helped to familiarise more people with the ideas and the technology. However, this local-level impact was still mostly on the margins. As the renewable energy programme developed, and large-scale projects emerged, its main emphasis was on corporate-level involvement. Although much of this was technology-led, professional engineers increasingly were 'on tap' rather than 'on top'. Perhaps inevitably, given the move to commercial projects, financial management issues moved to the fore, along with the need to engage in political lobbying to try to influence the direction of government policy. So it became less about the technology.

Within government circles, it was similar. Renewables had both supporters and detractors, with some Ministers being quite enthusiastic, others less so, but few had technical backgrounds. Civil Servants often played a key background role. Most were not technical, the classic generalists. But some did very well. For example, Godfrey Bevan, who was Director of the DTI's Renewable programme until his early retirement in 2001, was evidently well thought of by many in the renewable energy community. However, while in the early phase of the programme, as we have seen, the policies that emerged often reflected technical analysis, gradually economic and political factors and views began to dominate, and that was to become even more apparent post 2000.

5.2 Interim Policy Analysis

The successful development and deployment of new energy technologies seems vital, given the problems that had emerged with the existing range of options. However, as the government-defined UK renewable energy research programme outlined above illustrated, the management of R&D and the transition from R&D to commercialisation represented key problems in strategic terms.

To put it simply, there was a need for an assessment of which areas to support and then an assessment of when and if a technology has reached the end of its R&D phase and was ready for the next step, followed by a decision on what the next step might be and who should make it happen. Increasingly, in the 1990s, the UK government's belief was, firstly, that the bulk of the R&D had been completed and that industry should increasingly take over responsibility for any further R&D work that was needed, and, secondly, that it was up to the private sector to take any commercially attractive options on to full deployment, with the NFFO providing some interim support.

Unfortunately, given the economic and investment climate in the UK during the early 1990s especially, the main companies and investors in the private sector were, in general, more interested in what they saw as less risky energy projects, based on existing cheap fuels and more conventional technology, for example, natural gas used to generate electricity in cheap combined cycle gas turbines. The resultant 'dash for gas', in the main, effectively left only the smaller companies to explore the renewables, often using overseas equipment and capital, for example, Danish wind turbines and Japanese finance.

What support there had been from the private sector for renewables had, in the main, been focused on near-market NFFO projects, rather than on R&D and the development of the more advanced technologies. Energy Paper 55 projected that industry would take an increasing share of R&D, and that, by 1992–1993, the level of non-Departmental funding (i.e. from the European Community and from UK industry, but excluding that associated with NFFO projects) would rise to over £20 million p.a. In the event, it only reached around £8 million by 1991–1992 and fell off subsequently to around £6 million by 1992–1993 (NAO 1994).

Moreover, within the renewable part of the NFFO, with the exception of wind power, the emphasis was on relatively conventional projects. The majority of the projects supported under the first three NFFOs (more than 56% by capacity) involve waste combustion, sewage and landfill gas. As we have seen, the projects requiring further R&D had, in general, been left aside, with state support being cut, for example, for offshore wind, wave and tidal, and little industrial support for R&D in these areas emerging.

This might be seen as short sighted in strategic terms. While waste-to-energy projects and on-land wind farms clearly had a reasonable potential, there were local environmental impact and land-use limitations. Certainly, it has proved hard to get planning permission for many of the waste combustion projects, and, as noted earlier, some of the wind farm projects also met with opposition, mainly on visual intrusion grounds. Indeed, that has become a major issue, as has opposition to some biomass combustion projects, as we will be exploring in the next part of this book.

By contrast, offshore wind, wave and tidal stream energy represented large energy resources, with less potential for objections. For example, the offshore wind energy potential was estimated to be more than the UK's total electricity requirements (REAC 1992), and the exploitation of this offshore resource seemed likely to involve much less environmental impact, with minimal visual intrusion. It was the same for wave and tidal stream technology. Nevertheless, as indicated above, these options were seen by the DTI as relatively expensive, with the result that they had been relegated to the longer term.

Similar problems faced **photovoltaics**. There had been major state-supported PV cell programmes in the USA, Germany and Japan in the 1980s, and they were expanding. The UK, via BP and Shell's involvement, had a quite significant stake in the world PV market. But the UK government had not initially allocated significant funding to the PV, with the private sector basically being left to press ahead on its own and no UK market emerging. With a recession and competitive pressures building up in the global energy sector, BP and Shell's interest in PV subsequently declined. So the commercial prospects for PV in the UK were not looking too good. However, as noted earlier, following a re-assessment by ETSU in 1991, the UK government did provide some support for PV, and it was seen as a candidate for support under the then forthcoming fifth NFFO. The OST Foresight reviews had also been positive about PV, and some demonstration projects were supported for example, in schools, under the Government's £1 m Foresight Scholar programme. Even so, it was seen as relatively marginal.

The same was true for **fuel cells**, a hydrogen-to-power conversion technology first developed in the UK in the 1880s, with a practical 5 kW version being demonstrated in 1959 by Cambridge Professor Francis Bacon. Fuel cells had subsequently risen to prominence in relation to the US space programme, but were now being taken forward in the energy (and vehicle) context. The UK had continued its involvement, as one of the early pioneers, but with only limited government support. For example, the DTI provided £600,000 for work on fuel cells in 1992–1993, rising to £1.9 million for 1993–1994. However, with other countries (notably the USA and Japan) having significant fuel cell programmes, it was suggested that the UK might end up being what the 1994 ETSU overview report R83 called 'an informed customer' for fuel cell technologies bought in from overseas. That formulation seems quite defeatist, abandoning any 'first mover' advantage and indeed the UK's existing track record in fuel cell development. For example, it had supplied cells to NASA for use on the Apollo moon mission. However, there was the argument that the UK could not hope to cover all the options. It had to make choices.

The choice of technologies to focus on and the way they were developed in the 1980s and 1990s was clearly shaped by the various technical and strategic assessments reviewed above, but the assessments, and the support systems, were increasingly framed within a market context, in relation to short- to medium-term economic competitiveness, and the political aim of reducing state spending and expanding private sector involvement. However, there was also a wider international perspective. The UK had joined the European Economic Community in 1973. Although policy interactions in the energy area with the continental Europe were often not very significant, at least initially, as European integration strengthened in the 1980s and beyond, the influence of, and interaction with, what had become the European Union began to increase. See Box 5.2.

Box 5.2 The European Dimension

UK progress in renewable energy cannot be seen in isolation from developments in the rest of Europe. As time went on, the influence of the EU began to increase, although, while the UK did play a leading role in the EU input to the Kyoto climate change negotiations, to some extent the UK was seen as a bit of a backwater in terms of renewable energy development. It had a large resource, but was not developing it very rapidly.

A 1993 report for the European Commission 'The European Renewable Energy Study' (TERES) estimated that, at that time, the European Union as a whole only obtained around 4.3% of its energy from renewables. However, the report argued that, if full account of environmental concerns was taken on board, by 2010 Western Europe could, on the most ambitious scenario, be obtaining up to 13% of its primary energy from renewable sources. The pattern varied around Europe, with Spain being seen as having the potential to generate 20% of its energy from renewables, Italy 23%, but the UK only 9% by 2010.

The CEC TERES report explained the relatively poor UK showing as follows: 'Although the UK boasts good potential for exploitation of renewables resources, it is also the EC's most self-sufficient fossil fuel user. Therefore, a long term strategy for energy is not seen as major government priority. With the demise of the Department of Energy in 1992 went the last remnants of such a strategy.' It added 'The development of the country's full potential for renewable technologies can only be fostered within such a framework and thus is unlikely to be realised in the medium term, although the market imperfections introduced to favour renewables should allow the Government's modest aims to be met' (CEC 1993).

This might be thought to be a little harsh, in that the UK Government has prided itself on having a significant, but also very cost-effective, renewable energy programme. However, some other European countries had pressed ahead with more of a sense of urgency, and achieved some success. Indeed, by 2010 the (by then enlarged) EU was obtaining around 13% of its energy from renewables, much as TERES had projected just for the Western EU, but the UK has only reached 3.7%, well short of even the low 9% it predicted (Eurostat 2018).

In terms of electricity generation, by 1995, the EU as whole was generating 13.8% of its power from renewables and that rose to 14.7% in 2000. The UK figures were 2.3% and 3.2%, respectively, with the UK falling near the bottom of the EU league. An EU Directive, in 2001, set a target of getting 21% of EU electricity from renewables by 2010, with the UK being given, and agreeing to, a 10% target, well below most others. To be fair, unlike the UK, some EU countries did have the advantage of existing large hydro inputs, but even leaving that out, hydro-rich Finland, Portugal and Austria still had 2010 targets of over 20%, while Denmark, with no hydro, was aiming for 29% (EEA 2014).

The EU would help member countries to reach their targets, and, clearly, technologically as well as financially, there were advantages in the UK collaborating with European partners in many area, not least to catch up. There were already some collaborative energy projects and of course the EC provided pan-European support for R&D via the Joule/Framework scheme, and for commercial demonstration via the Thermie scheme. In addition, there was the Altener scheme, designed to provide organisational and institutional support. The prospects for fruitful technical and R&D collaboration certainly looked good (EUREC 1997).

We will be returning to look at comparative post-2000 EU and UK progress figures in more detail in Chap. 8, but sadly, as we shall see, in percentage terms, the UK has remained near the bottom of the EU renewable energy league table.

By the 1990s, the EU had become a major political force globally, promoting renewables as a response to climate change, with the UK joining in EU efforts to get international agreement on carbon emission reduction programmes, under the UN Framework Convention on Climate Change, at the climate summit held in Kyoto in 1992.

As we shall see, this process has continued, as did the influence of the EU on UK policies, for example, via the emergence, in the 2000s, of quite ambitious and rising pan-EU Renewables Directive targets. However, the EU-UK relationship has not always been an easy one. For example, as we shall see, the UK has struggled to meet the renewables targets set by the EU, and now, with BREXIT, for a range of reasons, divorce seems imminent, with uncertain impacts on UK energy policy.

Mention should also be made in this context of Scotland's aspirations for independence. Given that most of the UK's best wind resource is in Scotland, independence would have a major impact on the rump UK. Even without full political independence, Scotland does have a significant level of devolved planning powers, and, under a government led by the Scottish National Party (SNP), it has developed its own very radical energy policy, based heavily on renewables, wind especially, with the SNP Government opposing new nuclear plants. As we shall see, in stark contrast to the Westminster government's plans, the SNP Government looked to 100% of its electricity consumption being met from renewables by 2020 (Scottish Government 2009). Clearly, Scotland was setting the bar very high, and, as we shall see, it has succeeded in pressing ahead with renewables rapidly.

5.3 The UK Renewable Programme: Interim Summary

To round off this interim analysis of the UK renewables programme in the twentieth century, it is helpful to summarise some of the key issues and positions that have been revealed in the account above. One starting point is the scale of potential UK renewable resource. As we have seen, that had been much debated. But, in 1992, the Government's REAC estimated that renewable might, on a mid-range estimate, be able to supply up to 20% of the UK's electricity at competitive costs by 2025, and it put the total theoretical resource potential at more than three times UK requirements (REAC 1992). The DTI's 1994 Energy Paper 62 came to similar conclusions (DTI 1994). The 2025 estimates seem reasonable, if a little cautious, for the time. With hindsight, it had turned out to be possible to do much better—30% by 2018.

However, looking to further expansion, we can perhaps learn something from the approach that had been adopted in earlier phases reported on above. There might be problems in achieving major contributions from renewables, above 40–50%. Certainly the pattern of changing technological fortunes, priorities and policy reversals outlined above illustrates the difficulty of trying to 'pick winners' at the R&D stage. In the renewable field, £42 million was spent on the UK geothermal programme before it was wound up, £14 million on the tidal programme and £17 million was spent on one large wind turbine, the 3 MW unit on the Orkneys, a design which was not in the event taken up commercially.

Differing conclusions can be drawn from these outcomes. On the one hand, it could be argued that, in some cases, there had been a failure of nerve and a lack of commitment to following through, when initial results looked poor, coupled with erratic funding patterns due to shifts in policy.

As noted earlier, in its final report, in 1992, just before it was wound up following the demise of the DEn, the Select Committee on Energy commented that 'it is difficult to regard the history of renewable R & D funding in the UK as other than a history of volte faces, premature judgements and plain errors'. A little more charitably, it also suggested that 'the Department has attempted to establish costs at too early a stage in technological development' (Select Committee on Energy 1992).

Another view is that just 'throwing money' at projects was not necessarily likely to lead to success. A report on the UK renewable energy programme by the all-party House of Commons Public Accounts Committee published in 1994 was critical of the what it saw as the large sums spent on the large wind turbine and geothermal hot dry rock programmes. The key issue raised was that the R&D programme did not seem to have led directly to commercial success based on UK technology (PAC 1994).

A more general point was made in a 1994 report on 'the UK's renewable energy RD&D programme' by the National Audit Office, an independent government agency charged with carrying out 'value for money' reviews of government programmes. It was concerned by what it saw as a lack of emphasis on the export potential of UK renewable energy technology and by the focus on large projects, which it saw as 'reflecting the requirements of the pre-privatised electricity supply industry' which had 'led to a few projects and programmes receiving a large share of the £340m spent' (NAO 1994).

Finally, there is the view that breaking new ground is inevitably risky, and that, in fact, the programme had been reasonably successful: useful results had been produced and valuable experience gained, with the weaker options having been weeded out.

The UK government, in effect, took the latter view, arguing that the money had been well spent in identifying which of the renewable options were likely to be commercially viable. As a consequence, by 1994, it felt that its R&D programme had reached the point when it could be cut back, with the viable options being taken on by the private sector, with support from the NFFO, and with industry providing R&D support for any newly emerging options. Thus, Energy Paper 62 saw the NFFO as the primary mechanism for supporting the near-market options, with the DTI suggesting that it could help 'draw down' the £3 billion or so needed from industry if the government's 1500 MW by the year 2000 renewable target was to be met (DTI 1994).

While this process seemed to be going reasonably smoothly, if slowly (and leaving aside lack of commitment to R&D), the emphasis on commercial viability may sometimes have conflicted with environmental concerns. For example, wind developers were likely to seek the windier and more profitable upland sites, but these tend to be environmentally sensitive, and some projects fell afoul of local objections. Similarly, waste combustion was one of the most commercially attractive options, but there have been local concerns over emissions.

The emphasis on short-term commercial viability, and reliance primarily on the private sector to provide finance for R&D, presented the less-developed options with even more problems. It could be that, once the existing near-market options had been transferred to the market, the supply of new projects for the NFFO, or whatever followed it, by way of market enablement mechanism (e.g. the new Renewables Obligation (RO) system), would dry up. We will be returning to that issue in the next part of this book.

5.4 Interim Conclusions: Alternative Approaches

With concerns about greenhouse 'global warming' being taken increasingly seriously, some argued that a more robust approach would have to be adopted, namely the rapid development of the full range of renewables, as a key strategy for environmental protection and as part of a shift towards a sustainable energy future. That view seemed likely to spread.

Certainly renewable energy had been viewed in strategic terms in much of Western Europe, for example, in Austria, the Netherlands, Denmark and Germany. In these countries, initial direct capital investment by the state (e.g. via capital grants or rebates to developers) had been followed by market support follow-up schemes of an arguably somewhat more straight-forward kind than the NFFO, the so-called Feed-in Tariffs. Germany had been offering particularly large subsidies for wind. These subsidies were often justified on the basis of environmental benefits rather than on narrowly defined economic competitiveness.

This approach had been resisted by the UK government: instead competitiveness has been the watchword. As the DEn put it in 1988 in evidence to the House of Lords Select Committee on the European Community 'we do not support the proposition that renewables should be put in a preferential position so their utilisation comes forward in an uneconomic way' (House of Lords 1988).

The advent of the NFFO scheme did not change this position fundamentally: it was seen as a way to help renewables reach competitiveness, although obviously it did represent a form of subsidy, but also a contribution to reducing greenhouse gas emissions: by the year 2000, the projects supported under the renewable energy NFFO programme were expected to avoid the emission of around 2 million tonnes of carbon p.a., compared with the 10 million tonnes the UK has promised to cut from emissions following the Rio Earth Summit in 1992. Energy Paper 62 suggested that, in principle, renewables might save up to 38 mTC (or 24% of 1992 UK emissions) by 2025 (DTI 1994). But obviously this was unlikely to happen unless there was a concerted effort to develop renewables: arguably, it would need much more than just the NFFO.

Unfortunately, at least for renewables, the post-privatisation 'dash for gas' (i.e. the switch-over from coal to lower emission gas-fired power stations) had temporarily allowed the UK to avoid making any other major changes in the way power was generated and used. However, this respite was unlikely last indefinitely. Energy demand was expected to increase and overwhelm the emission savings that could be made by substituting gas combustion for coal combustion. In addition, gas reserves were not unlimited. If the market-led 'dash for gas' continued unabated, then UK gas reserves could be depleted quite rapidly and then gas would have to be imported, presumably from Russia, the other main options being Algeria or the Middle East. In either case, there could be problems with maintaining security of supply.

Issues like that began to be raised in the late 1990s and, as we shall see, became quite central in the 2000s, with the development of UK's shale gas resource being seen as a possible solution. That and the associated problems (increased emissions and local environmental impacts) were still in the future, but there were still plenty of reasons for developing indigenous renewable energy sources, and making a serious commitment to energy conservation. The other alternative was of course nuclear, though that was unpopular.

There was certainly some political support for a change in approach. While in opposition, in the 1980s and early 1990s, the Labour Party had indicated that it wanted renewable energy to expand to supply 10% of UK electricity by the year 2010, and it had also made a strong commitment to energy conservation. In 1986, in the wake of the Chernobyl disaster, the party Conference overwhelmingly confirmed its 1985 decision to not build any more nuclear power plants. When in power however, from 1997 onwards, it shifted approach, eventually backing nuclear, as did the Lib Dems, when in the coalition with the Conservatives. That was in the future, but it was in line with a common view, established in the pre-2000 period, that renewables were constantly being marginalised in preference to nuclear. Certainly there was strong support for renewables, and that was unlikely to change in the 2000s. However, it might have to be strengthened, if they were to move ahead.

To round off this interim review, it is worth looking at how that process might occur and who might be involved. Awareness of the benefits of renewables had been aided by the work of the various trade lobbies that had grown up, and they were clearly going to continue to play a role, both in lobbying government and in public outreach. Some of them had their roots in early enthusiast and professional organisations, like the initially academic-dominated British Wind Energy Association (BWEA), which was set up in 1979. As the wind industry expanded, that became more of an industrial lobby, and later expanded its coverage from wind to also include wave and tidal power, eventually (in 2009) changing its name to RenewableUK (RUK). The Renewable Energy Association (REA) was (and still is) the other main UK trade lobby group, covering all the renewable options, but focusing on solar and biomass, in conjunction with groups like the Solar Trade Association (STA) and the Anaerobic Digestion and Bioresources Association (ADBA). Interestingly, in parallel, the long-established industry-based Combined Heat and Power Association, led for many years by David Green, had often been a focus for critical thinking and lobbying on sustainable energy. It has now been reconfigured as the Association for Decentralised Energy, lobbying for a range of low-carbon and smart energy technologies at the systems level (ADE 2018).

Fragmentation by technology was sometimes a problem. Indeed, at one point, some in the wind lobby voiced concerns about too much attention being given to other renewable options (Renew 2001). Clearly, while some rivalries were perhaps inevitable, the government would naturally prefer the renewables sector to speak with one voice, and strategically that would make sense, but attempts to unify BWEA/RUK and the REA have so far not proved successful. However, they do try to co-ordinate their lobbying.

In terms of more indirect lobbying, over the years, many companies have supported PRASEG, the all-party Parliamentary Renewable and Sustainable Energy Group, with, typically, it having around 130 Parliamentary members. Set up initially in the mid-1970s as PARLIGAES, the Parliamentary Liaison Group for Alternative Energy Strategies, as should have been apparent from the story so far, it has provided a valuable forum for discussion of strategic issues. I should say that I was a founder member of PARLIGAES, with Labour back-bencher Frank Cook being one of its early chairs (PRASEG 2018).

Organisations like these, along with the various environmental pressure groups, notably Friends of the Earth, Greenpeace and WWF, have produced a steady stream of often-high-quality reports and studies, contributing to the debate over the direction of UK energy policy. Their influence seemed likely to continue. Certainly, in the 2000s, with concerns about climate change moving up the agenda, the environmental groups' message began to have more influence on overall policy, although, as we shall see, the practical focus in what then emerged was more on how to finance the expansion of renewables, an issue of obvious importance to the trade lobby groups.

References

ADE. (2018). *Association for Decentralised Energy Web Site*. http://www.theade.co.uk

Boyle, S., & Brown, P. (2003, May 31). Farmers Burned as Green Energy Plant Faces Export. *The Guardian*. https://www.theguardian.com/environment/2003/may/31/energy.politics

CEC. (1993). *The European Renewable Energy Study* (*TERES*). Report for the Commission of the European Communities, Brussels.

DTI. (1994). *New and Renewable Energy: Prospects for the UK*, Energy Paper 62. London: Department of Trade and Industry.

DTI. (1999). *New and Renewable Energy: Prospects for the 21st Century*, Consultation Paper. London: Department of Trade and Industry.

EEA. (2014). *European Environment Agency Data*. https://www.eea.europa.eu/data-and-maps/indicators/renewable-electricity-consumption-1/assessment-2

ETSU. (1993). *Tidal Stream Energy Review*. Energy Technology Support Unit, Report 05/00155, Harwell.

ETSU. (1999). *Energy Technology Support Unit Supplement, R 122, to the DTI*, 'New and Renewable Energy: Prospects for the 21st Century' Review.

EUREC. (1997). *The Future for Renewable Energy*. London: The Association of European Renewable Energy Research Centers, James and James.

Eurostat. (2018). *Share of Energy from Renewable Sources*, Eurostat Data. http://appsso.eurostat.ec.europa.eu/nui/show.do?dataset=nrg_ind_335a&lang=en

House of Lords. (1988). *Alternative Energy Sources*. House of Lords Select Committee on the European Communities, Session 1987–88, 16th Report.

NAO. (1994). *The Renewable Energy Research, Development and Demonstration Programme*. London: National Audit Office.

NATTA. (2000). *Local Renewables*. NATTA Conference Report, Network for Alternative Technology and Technology Assessment, Milton Keynes.

Nuclear Review. (1995). *The Prospects for Nuclear Power in the UK*, Nuclear Review Report, Cmnd 2860, London.

OST. (1999). *Energies from the Sea- Towards 2020*. A Marine Foresight Panel Report, Office of Science and Technology, Department for Trade and Industry, London.

Owen, G. (1995). *Energy Policy, the Government and the Energy Regulators: A Case Study of the Energy Saving Trust*. CSERGE Working Paper GGEC 95–35, Centre for Social and Economic Research on the Global Environment, UCL/UEA.

PAC. (1994). *The Renewable Energy Research, Development and Demonstration Programme*, Session 1993–94, 42nd Report, Public Accounts Committee, London.

Piterou, A., Shackley, S., & Upham, P. (2008). Project ARBRE: Lessons for Bio-Energy Developers and Policy-Makers. *Energy Policy, 36*(6), 2044–2050.

PRASEG. (2018). *The All-Party Parliamentary Renewable and Sustainable Energy Group*. Web Site: http://www.praseg.org.uk

REAC. (1992). *Renewable Energy Advisory Group Report to the DTI on Renewable Energy*, Energy Paper 60. London: Department of Trade and Industry.

Renew. (2001). *News Coverage and Analysis of the May 2001 Editorial in Wind Power Monthly*, Renew 133 Sept-Oct, Renew On-line 33. http://eeru.open.ac.uk/natta/renewonline/rol33/2.html

Ross, D. (1979). *Energy from the Waves*. London: Pergamon.

Ross, D. (1995). *Power from the Waves*. Oxford: Oxford University Press.

Select Committee on Energy. (1992). *Renewable Energy*. Session 1991–92, Fourth Report, London.

Scottish Government. (2009). *Scottish Renewables Action Plan*. http://www.gov.scot/Publications/2009/07/06095830/0

6

The Market Takes Off *Into the 2000s*

6.1 A New Emphasis on Sustainability

Concerns about the impacts of climate change had been building for some time, but in the 2000s they deepened and led to renewed commitments to environmental sustainability and sustainable energy, with the UK playing a role, along with the EU. In 2000, Labour's deputy Prime Minister John Prescott, who had been centrally involved with the Kyoto negotiations, set up a UK Sustainable Development Commission and, in 2001, the Labour Government introduced a Climate Change Levy (CCL) system, which imposed a tax on industrial emissions, although with some exemptions. It was hoped that would stimulate companies to develop and use renewable sources.

In 2005, the Kyoto climate protocol came into force globally, and specific programmes for carbon reduction began to emerge in the EU. In 2009, the European Commission produced a Renewables Directive calling for member states to help the EU get 20% of its *energy* from renewables by 2020, building on the earlier target, mentioned earlier, of getting 21% of its *electricity* from renewables by 2010 (EC 2009).

D. Elliott, *Renewable Energy in the UK*, Energy, Climate and the Environment,
https://doi.org/10.1007/978-3-030-04765-8_6

The UK government had acceded to these targets, and to the wider EU and global carbon reduction targets. It also set national UK carbon targets, via the Climate Change Act, introduced by the New Labour Government in 2008, with a legally binding target of an 80% reduction (from 1990 levels) in greenhouse gas emissions by 2050, and five-yearly carbon budgets (CCA 2008). In 2008, Labour also established a Department of Energy and Climate Change (DECC) to oversee the programme. A Renewables Advisory Board was also established, with 10 members from government departments and 16 from industry. In 2008, Labour also introduced proposals for an FiT scheme for small renewable projects. Looking more broadly, a Carbon Trust was also established, as a not-for-profit company, with its aims, as defined by the Prime Minister Tony Blair, being to 'take the lead on low carbon technology and innovation in this country, and put Britain in the lead internationally', for example, via a new Low Carbon Innovation Programme.

Clearly, climate change and the pursuit of sustainable energy/low-carbon options as a response had arrived centre stage. It is not my intention to explore the complexities of climate and sustainable development policy, nationally or globally, since it has been covered extensively elsewhere, including in the series of studies of which this book is part (Elliott and Wood 2018). Instead, my aim here is to look at what the UK has done in practice with renewables, and how it was done. However, it has to be noted that renewables were not the only option for carbon emission reduction, and the pursuit of other options, nuclear in particular, has inevitably shaped, and limited, the renewables programme, as we have already seen. That influence continued in the 2000s. Indeed, if anything, it became stronger.

Although I follow that through in what follows, I have not explored the nuclear issue in any detail. That has been covered extensively elsewhere, including in my own studies (Elliott 2003, 2010, 2017). Neither do I look in any detail at the other main contender, Carbon Capture and Storage (CCS), which in theory offers the prospect of allowing for the continued use of fossil fuel. I have also reviewed that elsewhere (Elliott 2018). Suffice it to say that, quite apart from their many short- to medium-term problems, neither of these options seems to offer a long-term sustainable solution. Fast Breeder reactors might help extend the use of fissile fuel reserves, but like fossil fuel, uranium and thorium reserves are finite, and nuclear fusion maybe apart, renewables remain the only long-term sustainable energy supply options, with nuclear and CCS only offering, at best, a temporary reprieve (Elliott 2015).

Nevertheless, the account of the next phase of UK renewable development that follows inevitably reports on the impact of the continuing pressure from the still powerful nuclear lobby. The CCS lobby, by contrast, was relatively weak and played little part in this story, although the wider fossil fuel lobby certainly did, and still does, play a role in the wider battle for a sustainable future, with, in the UK, shale gas fracking becoming a major issue.

While, as we shall see, although fossil and, especially, nuclear influences clearly continued to play roles, and the issue of energy saving continued to be raised (but sadly mostly ignored), the main focus in the next part of the story was on the expansion of renewables. With the pace picking up, the focus in this part of the book moves on from general overviews and resource assessment activities, to specific funding programmes and deployment policies, with costs being key issue. That is not to say that supply-side technology and resource-scale issues did not still play an important role in the debate. For example, wave power re-emerged, following the review mentioned earlier of marine renewables under the TF programme, and the DTI decision to restart some funding. They were clearly under some pressure, with, in 2001, a Select Committee concluding that 'given the UK's abundant natural wave and tidal resource, it is extremely regrettable and surprising that the development of wave and tidal energy technologies has received so little support from the Government' (Science and Technology Committee 2001). See Box 6.1.

Box 6.1 Marine Renewables: Back from the Grave?

In its 2001 review of marine renewables, the House of Commons Science and Technology Select Committee came out firmly in favour of wave and tidal stream energy. It said: 'The current level of public spending is insufficient to give the technology the impetus it needs to develop fully...We recommend that the Government increase the amount of funding available for full-scale wave and tidal energy prototypes to prove the concept at a realistic scale.' It added: 'We recommend that a significant proportion of the extra £100 million of funding for renewables, recently announced by the Prime Minister, be made available for wave and tidal energy demonstration models' (Science and Technology Committee 2001).

At the public release of the report, the committee's acting chairman, Dr Desmond Turner MP, said that the Government should spend 'something of the order of £10 million' to start with, and then tens of millions a year, adding 'even if the Government should spend several hundred millions, the potential advantages to the British economy mean that it will be a very sound investment'. Noting that the potential wave and tidal resource was vast, during a Parliamentary debate on the issue on 5 April 2001, Turner argued that 'it is extraordinary that we should be funding wave and tidal power in inverse proportion to its availability'.

So some obviously felt a rethink was called for, this view being buttressed by the admission from the DTI that errors had been made in earlier assessments. During the Select committee hearings in March 2001, the civil servant head of the Government's renewables programme, John Doddrell, had admitted that the Government had blundered in 1994 when it finally abandoned its already cut-back wave energy programme. He said 'The decision to discontinue was taken on economic grounds. With hindsight, it was clearly a mistake', and he read out a formal DTI statement which said 'The decision was taken in the light of the best independent advice available. With the benefit of hindsight, that decision to end the programme was clearly a mistake' (Ross 2002).

That was quite an admission. However, all was not lost. As was pointed out, the DTI had in fact restarted the wave programme in 1999, and there were seven wave-related projects underway, along with one tidal stream project, to a value of £1.86 million, with the DTI contribution being £1.27 million. In addition, the ESRC had ten wave energy and tidal stream projects in place, to the total value of just over £1.1 million. It was noted that the DTI-supported projects involved 'the further development of existing design concepts, research to tackle key development issues and monitor prototype devices'. So they were mostly generic studies rather than new devices. But a further round of R&D funding was expected (a £19 million private-public package emerged in 2003). The Scottish Government was also offering £13 million support. The early and mid-2000s certainly saw many tidal stream projects being tested, along with some wave energy projects (Elliott 2007; NATTA 2007).

However, while debates continued on these and other technologies, costs were now an increasingly key issue. To try to establish what they might be in the period up to 2020, in 2001, the Labour Government initiated an Energy Review, carried out by the UK Cabinet Office's

Performance and Innovation Unit (PIU), with extensive inputs from academics contracted to work with the PIU team. The study made use of 'learning curve' analysis to try to identify trends in price reductions. If prices at successive stages in the innovation process are plotted against cumulative production volume on a log-log scale (i.e. with the data in both axes presented in logarithmic terms), then, in many cases, a straight line results. The slope varies with the technology, but the range is not great: gradients of between 10% and 20% are typical. Not all the energy options reviewed by the PIU could be assessed in this way, since, for the newer options, like wave and tidal current technology, the data was not yet available, so parametric engineering assessments and proxy assessments had to be used (PIU 2002).

Interestingly, the PIU found that, despite very large-scale funding over a long period, the nuclear learning curve slope was only 5.8%, which they attributed to the fact that it was a 'mature' technology, involving large, inflexible projects with long lead times. They also argued that the frequent emergence of completely new designs meant that there was less technological continuity, less opportunity for economies of production scale and less opportunity for learning. It did seem that nuclear power was an exception to the norm. By contrast, the results for some of the key new renewables were much more encouraging: the learning curve slopes for PV solar and wind were put at between 18% and 20%.

The final conclusions for the energy options reviewed by the PIU are shown in Table 6.1. Clearly, these long-term estimates were speculative and relied on a range of assumptions about policy developments. For example, if funding was not provided for new renewables or new nuclear technologies, then the picture could look very different. But the simple message from Table 6.1 was that, although some of the renewable energy options were relatively expensive at present, they were likely to get cheaper, as would Combined Heat and Power/cogeneration (CHP). By 2020, most of these options would be cheaper than (advanced) coal and nuclear, and some would be cheaper than gas. On that basis, it seemed clear which options should be favoured, with wind an obvious leader.

Overall, the PIU concluded that 'the immediate priorities of energy policy are likely to be most cost-effectively served by promoting energy efficiency and expanding the role of renewables', and it called for 'step changes in

Table 6.1 Cost of electricity in the UK in 2020 p/kWh (PIU)

On-land wind	1.5–2.5
Offshore wind	2–3
Energy crops	2.5–4
Wave and tidal power	3–6
PV solar	10–16
Gas CCGT	2–2.3
Large CHP/cogeneration	Under 2p
Micro-CHP	2.3–3.5
Coal (IGCC)	3–3.5
Fossil generation with sequestration	3–4.5
Nuclear	3–4

Source: PIU (2002)
Note the very high estimate for PV costs that has turned out to be wrong

energy efficiency and vehicle efficiency' and for the target for the proportion of electricity generated from renewable sources to be increased to 20% by 2020. Crucially, it noted that 'there is no current case for public support for the existing generation of nuclear technology' although, no doubt seeking to avoid undue conflicts, it added 'there are however good grounds for taking a positive stance to keeping the nuclear option open' (PIU 2002).

In 2003, the Labour Government duly produced a White Paper on Energy which, based on the PIU's work, said that the current economics of nuclear power 'make it an unattractive option and there are also important issues of nuclear waste to be resolved'. Instead, it supported renewables and energy conservation (DTI 2003).

As we shall see, the Labour Government subsequently changed its view on nuclear, but, for the moment, with that issue off the agenda, the focus then moved on to how to support a range of renewables, with wind power featuring strongly. There were also some demand-side/energy-saving initiatives, such as the Zero Carbon Homes policy, which we will look at in Chap. 7. On the supply side, as well as funding and cost issues, the rapid expansion of renewables like wind also introduced some new issues in relation to their environmental sustainability. Rather than the issue of the *scale* of the resource that had dominated some of the earlier debates, a key issue was the social and environmental *impact* of using it.

As noted earlier, opposition to some wind projects had emerged in the early phase (Elliott 1994), but, as the programme expanded, the issue of impacts loomed larger, especially for those opposed to the programme overall. That had political implications. Much of the local opposition to wind farms came from traditionally Conservative rural areas, the so-called Tory Shires, and that shaped Conservative Party policy, eventually, after it took power again in 2010, leading to a block to on-land wind projects. We will be looking at that later. However, while local opposition to wind projects was apparent in some areas, in general public reactions to renewables were very favourable, with typically 80–85% of the public supporting them, and support actually rising over time, including for onshore wind. See Box 6.2.

Box 6.2 Public Attitudes on Renewables

The government has been producing quarterly public opinion surveys of views on energy since 2012. The 'wave 23' survey in 2017 found 82% supported renewables overall. Opposition to renewables remained very low, at 3%. PV was the most popular renewable, with 84% support, while 79% backed offshore wind, wave and tidal, 74% supported onshore wind and 69% supported biomass. That contrasts with the 33% who backed nuclear energy. Going back to 2012, the 'Wave 1' results that year showed that, of the sample, the highest level of support was for solar (83%), offshore wind (76%) and wave and tidal (75%), and 66% supported onshore wind. So support has gone up for all renewables. Indeed, in the most recent (2018) survey (Wave 25), it had reached an all-time high level, with solar at 87% and offshore wind at 83%. Wave and tidal got support from 81%, while onshore wind got 76%, and (unchanged since 2017) biomass got 69%. Support in 2018 for renewables overall among UK residents had climbed to 85%, its highest level since the government first began recording attitudes towards energy and climate change issues, contrasting with the strong public opposition to and low support for shale gas fracking, with support for it reaching an all-time low, at 13%, in the 2017 BEIS survey (DECC 2012a; BEIS 2017a, 2018a).

Looking further back, support for renewables, and also for energy efficiency, seems to have been consistent and high over time. A 2002 National Opinion Poll carried out for the Energy Saving Trust found that 76% believed the Government should invest time and money developing new ways to reduce energy consumption, 85% wanted government investment in 'eco-friendly' renewable energy (solar, wind and water power) and only 10% said the government should invest time and money in building new nuclear plants (Elliott 2003). For some further data and views on wind power, see the DTI's 'Wind power: 10 myths explained' (DTI 2005).

The other key new issue was the *variability* of the renewable resource. Wind and solar inputs are weather dependent, and the problem of the resultant 'intermittency' of supplies had long been pointed to by those opposed to the wide-scale use of renewables; they were not a reliable option. With small amounts of renewable capacity on the grid, the variability of the sources did not matter too much, and there were ways to deal with it. However, as renewables expanded, as we shall see in the next chapter, this issue became more central, not least since dealing with this variability added costs. Clearly, for this and other reasons, a shift to sustainable energy did come with costs, and the economic issues proved to be central during the next phase. As a result, it was typified by attempts to come up with new financial support mechanisms designed to keep prices down.

6.2 The RO Story and the UK FiT

In terms of the development of renewables, as noted earlier, the main mechanism for ensuring expansion favoured by the Labour Government was the RO, which it introduced in 2002, replacing the NFFO. Like the NFFO, it was a market-based system, but did not use price/capacity auctions. Electricity supply companies were awarded ROCs for each MWh of eligible renewable electricity they bought from generators, set against government targets. If they failed to meet the target, they would incur a financial penalty, although they could avoid that by buying in any spare ROCs available from suppliers who had managed to exceed their targets. So the ROCs had a tradable value. There would be an incentive to invest in sufficient renewable capacity so as to avoid having to buy ROCs, or pay the penalty, but also, depending on the market values of the ROCs, an incentive to earn more than they needed, so that they could be sold on.

The level at which the penalty surcharge was set would, in effect, set the point at which it became worth buying in ROCs (it was initially set at 3 p/MWh, rising later in stages to 4.5 p), but the level of capacity needed would gradually rise as the Obligation targets were increased. Electricity supply companies were initially required to move in steps to

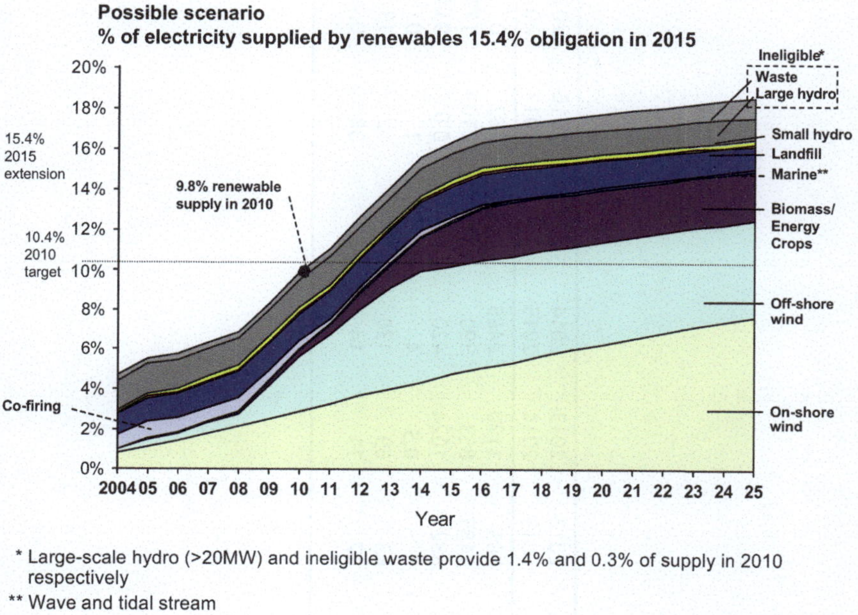

Fig. 6.1 DTI Outline projections, based on OXERA and ILEX data. (DTI 2004)

source 10.4% of their electricity from renewable generators by 2011, the national target having been set, in line with EU agreements, at 10% (of electricity) by 2010. In theory then, the RO quota/trading system would thus ensure that renewables would expand smoothly, with the extra cost being passed on to the consumers, typically adding 5–6% to their power bills in the early stages. A longer-term output target of 15% of electricity by 2015 was set, along with an expectation of near 20% by 2020. See Fig. 6.1 for an early DTI indicative scenario.

The system proved to be reasonably successful in building up some capacity, with wind and biomass projects (mostly using landfill gas and sewage gas) going ahead in the initial phase (see Table 6.2 later), but there were some problems. While landfill gas and sewage gas were relatively economically attractive options, as were some onshore wind farms and hydro projects, other so far less-developed renewables, and smaller companies and projects, were not able to benefit. It was the large suppliers and projects that won out. Moreover, some mature wind projects on

Table 6.2 Renewables Obligation accredited capacity

	2007–2008	2008–2009	2009–2010	2010–2011	2011–2012	2012–2013	2013–2014	2014–2015	2015–2016	2016–2017	2017–2018
Solar PV	1	0.4	0.3	0.3	6	285	1170	2485	1233	662	0
Onshore wind	694	603	495	499	1231	1485	853	390	739	2109	173
Offshore wind	190	90	352	1063	645	882	1024	420	50	0.0	914
Landfill gas	37	28	22	32	0	8	0.5	4	0	1	0
Biomass fuelled	33	13	70	278	60	63	69	130	59	42	13
Others (hydro, sewage gas, tidal)	15	124	11	15	1	12	11	14	3	30	0.2

Ofgem (2018)

good sites might get more support than they needed. Similarly for some of the low-cost biogas projects and for hydro. In 2005, OFGEM, the market regulator, claimed that, over the period 2002–2005, around £750 million may have been paid out in excess subsidy over and above what was needed by the biogas, hydro and wind projects, landfill gas being the worst case (Ofgem 2005). The National Audit Office calculated that, unless there were changes, by 2026–2027 around a third of the funding would be going to projects that did not need it (NAO 2005).

Subsequently, to avoid this, following a new energy review (DTI 2006a), a series of technology bands was introduced, with differing levels of ROC payment for each technology. For example, support for existing landfill gas projects was reduced to 0.25 ROCs/MWh and later to zero, while new landfill gas projects got 0.2/ROCs/MWh. Sewage gas was cut to 0.5 ROCs/MWh, and hydro (outside Scotland) to 0.7 ROCs/MWh. Onshore wind also fell to 0.7 ROCs/MWh. Efforts were also made to adjust the system to provide support for the less-developed options like wave and tidal power, by offering them multiple ROC allocations per MWh. The allocations varied, but, at one stage, tidal steam projects were being offered 2 ROCs/MWh and wave projects 3 ROCs/MWh, later expanded to 5 ROCs for small wave or tidal projects. Recourse was also made to grant aiding, to try to keep things moving forward, including, at one stage, £400 million for wind project support and £50 million for wave and tidal projects, on top of support via the R&D programme, which was running at £20–30 million p.a.

Some critics felt that these extra capital grant subsides (totalling £500 million or more) indicated that the market-based ROC system was basically not fit for purpose. To get enough new capacity built, extra support, almost equal to or even more than the amounts available via the ROC system, was having to be provided. In its first two years, the RO only delivered support worth £282 million (2002–2003) and £405 million (2003–2004). Even so, these subsides, combined with the grants, were significantly larger than had been provided to renewables by the NFFO, which, in its last few years (see Table 4.2), was only yielding around £55–65 million p.a. in total (Wilson 2001).

However, some critics felt that the level of capacity growth that was resulting from the RO was insufficient, and some also felt the cost to consumers was high, as were the overall costs of the power delivered, compared to what was emerging from the guaranteed-price renewable energy Feed-in Tariff system (REFIT) used in Germany and elsewhere (Toke 2007a). Certainly, the cost of electricity was initially quite high. For example, in 2003, the RO-supported wind projects were generating electricity at around 9.6 euro cents/kWh, whereas the equivalent figures for Germany ranged from 6.6 to 8.8 euro cents, while for Spain the cost was only 6.4 euro cent/kWh (Grotz and Fouquet 2005). This seemed odd, since the UK has the EU's best wind regime, with generally higher wind speeds. Moreover, despite this advantage, the RO also seemed to be delivering much less capacity than elsewhere. For example, by 2005, only around 2 GW of wind capacity had been installed in the UK, compared to around 20 GW in Germany.

The cost pass-through to consumers under the RO for this capacity also seemed to be much higher than what was being charged to consumers in Germany under their REFIT system. For example, in 2005–2006, the UK's RO system cost consumers 3.2 p/kWh, whereas, in 2006, the German REFiT only cost consumers 2.6 p/kWh, despite its poorer wind regime and also supporting the installation of increasing amounts of much more expensive PV solar capacity (Ernst and Young 2008). Leading critic David Toke claimed that 'the current levels of renewable deployment have been achieved at a cost to the consumer that is over 40% higher compared to what could have been achieved with a REFIT organised in a way that is broadly similar to that operating in Germany' (Toke 2007b).

Part of the problem was that, under the RO, the income to the power companies varied, depending not just on the market for electricity but also on the ROC market. The later could vary dramatically and unpredictably. That uncertainty made it hard for developers to get investment capital for new projects at low rates, so borrowing costs were higher and the suppliers had to charge consumers more, and fewer projects went ahead than in Germany. The financial risks were too high (Mitchell et al. 2006).

There were other problems with the RO system. The competitive pressures the RO created meant that companies tended to seek out the more profitable high wind speed sites, which were usually of more scenic value. This added to the backlash from preservation groups and local interests.

At one stage, 70% of proposed wind projects were being opposed, although some of these got through on appeal. In Germany and (initially) Denmark, under FiT systems, most wind projects were on sites with much lower wind speeds than would be considered economic in the UK under the RO, and there had been very little local opposition to them. Indeed, since most of the initial projects in Denmark, and in Germany later on, were locally owned, they were usually strongly supported and even sought for. For example, the wind co-op enthusiasts often quoted the old Danish proverb 'Your own pigs don't smell.'

By contrast, almost all the wind projects in the UK were owned by large companies and often financed with overseas investment. They could be portrayed as being imposed on unwilling communities in inappropriate locations by profit-seeking outsiders. That may be an overstatement, but certainly, under the RO system, it was very hard for local wind co-ops and small projects to get started. Only two managed to get going.

Basically, the RO was for the 'big boys' and for established large-scale renewables. It certainly worked well for them. Once built, these projects were relatively profitable. The National Audit Office noted that the internal rates' returns were similar to wind projects elsewhere in the EU (NAO 2005). That may be one reason why the wind industry was happy with the RO, even if it did not lead to much capacity being installed.

That is not to say that the RO did not deliver some capacity, with around 5.5 GW of onshore wind by 2010, and the beginnings of an offshore wind programme. As we shall see, that blossomed subsequently. As is indicated in Fig. 6.1, biomass energy crops were also seen as a significant option, possibly involving short rotation coppicing, but for the moment the emphasis was on waste combustion, including municipal solid waste, and biogas from sewage plants and landfill sites, which although, as we have seen, squeezed out later by the adoption of Technology Banding, initially made up the largest share of the overall renewable power output. Table 6.2 shows the capacity contracted in each year over the period 2007–2018, up to the point when the RO was replaced and closed to new projects.

As can be seen, onshore wind initially dominated, but was soon followed by offshore wind. Landfill gas gradually tailed off to zero. By contrast, PV solar started low but accelerated ahead, reaching a peak of almost 2.5 GW in 2014–2015, after which new projects over 5 MW were blocked from the RO. We will be looking at that later: it was part of the policy of limiting consumer cost pass-through and halting the spread of large solar farms. New onshore wind projects were also blocked from the RO from 2017, on the basis of new policies, after hitting a peak of over 2 GW the year before. We will return to that later.

While large solar PV projects had gradually done well under the RO, until they were blocked, the RO was not suited to smaller PV projects. Lobbying for alternative forms of support for small domestic-scale PV intensified, especially given its growth in Germany under its FiT system. In 2000, an attempt was made to establish a 'net metering' system to support PV take-up, but it was rebuffed (Renew 2000). In part, that was because the RO was not the only support on offer for PV or other small-scale renewables. In 2006, to replace the earlier £31 million Major PV Demonstration Programme and the £10 million Clear Skies small-scale renewables programme, the Labour Government introduced a £80 million Low Carbon Building Programme (LCBP). This was mostly for large public schemes, but with £6.5 million set aside to stimulate the uptake of domestic-scale micro-generation technology, including roof-top solar (heat and PV power) and micro-CHP heat and power units, and also, a little exotically, micro-wind. See Table 6.3.

There was some enthusiasm for the idea of 'self-generation' by users, as was beginning to happen in Germany with PV solar. However, the LCBP faced an overspend problem. Demand for domestic grants rapidly

Table 6.3 Micro-generation in the UK

No. of installations by 2006	
Micro-wind	650
Micro-hydro	90
Ground source heat pumps	546
Biomass boilers (pellets)	150
Solar water heating	78,470
Solar PV	1301
Micro-CHP	990
Fuel cells	5
Total	82,202

DTI (2006b)

outstripped the money available, even with a monthly cap on allocations imposed. The scheme was temporarily suspended in April 2007, but restarted with an £6.2 million extra provided to allow it to run up to 2008. There was clearly a large unmet need, as was also indicated by demand for support for local small-scale projects under the £50 million two-year Community Energy programme that had been set up in 2002, and then extended to three years in 2004. Demand for support for small-scale renewables was clearly growing, but some felt a grant-aid approach was not the best way to meet the growing demand.

The lobbying pressure to set up an FiT system for small projects like domestic PV increased, and as noted above, this was finally acceded to in 2008, although the first contracts did not emerge until 2010, with funding for the LCBP continuing meanwhile. Once it got going, the UK FiT scheme, originally called 'clean energy cash back', proved quite popular with householders investing in PV on their rooftops, and also helped some community groups to set up local projects. See Box 6.3. However, the UK FiT scheme, unlike its German counterpart, was limited to small projects (under 5 MW) and was not expected to deliver much total capacity. It was seen as supplemental to the RO. That was still considered by DECC, which by this stage had taken over from the DTI, to be the main vehicle for support for most projects, especially large projects, such as wind farms and also larger PV projects.

Box 6.3 The UK Feed-in Tariff System

The UK FiT had two elements. Eligible consumers were given a fixed price for the renewable electricity they generated and used, though varying by the technology, but could also get paid for any power they exported to the grid, the export tariff initially being set a 3 p/kWh. The generation tariffs were pre-set and fixed for a period ahead, but (as in Germany) were subject to predetermined annual price 'degression', that is, a percentage cut-back, based on expected reductions in cost as the market built and the technology improved. The price degression rate for solar PV was set at 7% p.a., and 3% for wind, apart from systems under 1.5 kW for which it was 4%, all starting from 2012, but with an extra uplift of 0.5% for PV after 2015.

For the generation tariff, for the first period (2010–2011) it was proposed that PV solar generators would get 36.5 p/kWh for systems up to 4 kW and 28 p/kWh for systems up to 10 kW, while wind projects would get 26.7 p/kWh for wind turbines between 1.5 kW and 15 kW, 34.5 p/kWh for units up to 1.5 kW. Generation tariffs were also offered for small hydro (4.5–19.9 p/kWh depending on size) and also for AD biogas use (9–11.5 p/kWh), with no price degression (DECC 2010a).

It was expected that the FiT might yield returns on investment of 5–8% p.a. and lead to a cumulative 6 TWh of production (1.6% of the UK total) by 2020. Recipients of grants under the LCBP could also get FiTs while the LCBP scheme lasted, so, all in all, the package, including the export tariff, was quite generous. Rough independent estimates suggested that a (quite large) 15 kW wind turbine at a good site could get a 12% rate of return, but PV solar was more marginal at 7% p.a., although still a good investment. But, as with the RO and the NFFO, the cost of the FiT would be passed on to all consumers via their power bills.

The battle to get the FiT established had been quite hardly fought. The government seemed wedded to the RO and initially tried to limit the FiT to projects under 3 MW or less, to protect the RO, whereas the campaigners tried to push for 10 MW. 5 MW ended up as the compromise (Renew 2009). The Low Carbon Building Programme funding, enhanced with an extra £45 million in 2009, in effect filled in while the FiT was being set up.

While the FiT was aimed at PV and other mostly domestic-scale projects, and did quite well, wave and tidal power projects were eligible for ROCs under the RO, but did not develop much beyond continued testing of prototypes, for example, at EMEC, the European Marine Energy Centre, in the Orkneys, which was set up in 2003. Nevertheless, one project, Marine Current Turbines' 1.3 MW Seagen, did go further. It was installed in Strangford Lough in Northern Ireland and was able to make use of RO support. It got 2 ROCs/MWh. But no other projects were able to get RO support at this stage. That was despite an attempt to provide intermediate funding to help other wave and tidal projects become eligible for RO support. A Marine Renewables Deployment Fund (MRDF) was set up in 2004, offering £42 million for tidal current turbine and wave energy projects. However, it required projects to have proven themselves at sea at commercial scale for three months, and, in practice, no one was likely to build something speculatively in the hope of obtaining funding later, and no one applied. Energy Minister Malcolm Wicks admitted that 'because no wave or tidal-stream energy technology has yet accumulated three months of continuous operation, which is a key eligibility criterion of the scheme, all of this budget remains unspent' (Wicks 2008).

In 2008, the Renewables Advisory Board rather lamely said 'The MRDF is fundamentally a sound scheme. It, in itself, is not a failure, but the R&D process has failed to supply the technologies that the MRDF was established to support' (RAB 2008). So the scheme was fine, but the technology wasn't ready! The government then introduced an interim £22 million Marine Renewables *Proving* Fund, to make projects viable for the MRDF, and then the RO. Six projects did apply for this, and some got funds for development work (including the Pelamis and Oyster wave devices), but, in the event, only one new project (for part of the Meygen tidal stream scheme) eventually managed to get RO support, in 2017. Clearly, it was hard to get these new technologies moving via the RO.

The government was certainly trying hard to promote renewables though schemes like this and others, but arguably, it was hamstrung by the structure and framing of the RO. Evidently, it was not suited to emergent technologies. By 2010, the UK had around 10 GW of renewable capacity in place, roughly equal to its nuclear capacity, although delivering less output, given its lower load factors. Good progress, but not stunning. Well before that date, with the climate change issue pressing on the government and energy security also a concern, it was perhaps inevitable that Labour had looked back to nuclear.

Since this had a significant impact on subsequent developments, it is worth backtracking to look at what happened. As may be recalled, Labour's 2003 White paper had concluded that nuclear was unattractive, but its new energy review, led by Energy Minister Malcolm Wicks, came to different conclusions. It was carried out in 2006 by the DTI, which produced a consultation paper, citing changed energy security circumstances as one reason for the rethink. The final energy review report, which also had suggestions for revamping the RO, noted that 'the Government believes that nuclear has a role to play in the future UK generating mix alongside other low carbon generation options' and promised a White Paper outlining proposals (DTI 2006a).

This new position was not without its parliamentary critics. For example, the Environmental Audit Committee had made its views clear during the consultation phase: 'We remain convinced that the vision contained in the [2003] White Paper – with its focus on energy efficiency and renewables as cornerstones of a future sustainable energy policy – remains correct'

(EAC 2006). The government's advisory Sustainable Development Commission, which had just produced its own review of nuclear, was also not impressed.

A broad ranging DTI White Paper on Energy in 2007 (DTI 2007), which laid out plans for renewables and the RO, also repeated the commitment to making a final decision on nuclear. In early 2008, by which time the DTI had been replaced by the Department of Business Energy and Regulatory Reform (prior to the takeover by DECC), the Nuclear White paper emerged, commenting that 'new nuclear power stations would make a significant contribution to tackling climate change' and opening the way for a new private sector-led programme (BERR 2008).

There has been much debate over exactly what had happened and the influences that dominated, but the government's then Chief Scientific Advisor, Professor Sir David King, who was widely seen as being pro-nuclear, may have played a key role. He subsequently led a study which included a range of high nuclear scenarios (Butler et al. 2012). Whatever the reasons, certainly Prime Minister Tony Blair had adopted a pro-nuclear line. He famously said, during the earlier review phase, that nuclear was back on the agenda 'with a vengeance', although in the actual quote, he also referred to 'a big push on renewables and a step change on energy efficiency' (Blair 2006). His replacement, Gordon Brown, who oversaw the completion of the 2008 Nuclear White Paper, was similarly pro-nuclear.

It was not clear exactly how nuclear would be supported under the new policy, presumably not via the RO (although there were hints about that), but it and renewables would both be backed in future. This policy was also eventually adopted (after initial objections from the Lib Dems) by the incoming Conservative-Liberal Democrat coalition government, which took over in 2010. In addition, it introduced a raft of reforms and revisions, including a replacement for the RO, and also carried out new reviews of renewables, produced by DECC, the results of the first being published in 2011, as a 'Renewable Roadmap', with an update in 2012, followed by more reviews in 2013 and 2014.

Before we move on to look at the new government's policy shifts, these reviews, and a parallel Select Committee marine review, provide a convenient scene-setting framework for what then occurred, with a comparison of their results offering an interesting picture of how assessments changed over time. For example, initially still marginalised in the 2011 review, PV solar was subsequently re-assessed (in a 2013 review and then 2014 updates) to be a front runner. See Box 6.4. That outcome, and the other changing assessments and special pleading (e.g. on wave and tidal power), had implications for the new support programme that the new government introduced to replace the RO, and as we shall see, also for subsequent policy developments. The new government was having to deal with a rapidly changing technological situation, while at the same time making radical changes in the support system, as well as in policy.

Box 6.4 DECC Renewables Overviews and Another Marine Review

In its 2011 'Renewable Roadmap', the Department of Energy and Climate Change outlined how it saw each key option. It selected 8 technologies as likely to be key to meeting the UK's 2020 renewables targets, or offering potential for later. They included on- and offshore wind, marine energy, biomass electricity and heat, and heat pumps. PV solar was not amongst them (DECC 2011a). PV was given slightly more prominence in the 2012 Roadmap update (DECC 2012c). However, in 2013 DECC produced an upbeat solar road map (DECC 2013a) and then a 2014 update (DECC 2014a), by which time PV had reached 2.7 GW.

So, at this stage, after initially having seen PV as a relatively limited option, DECC was in promotion mode, although pushing PV as a technology that would hopefully soon no longer need government support. In its subsequent 2014 review of 'options for future support', DECC said, 'we consider solar PV now to be an established technology in the UK', and this was clearly true—it was booming. Crucially, DECC added 'Solar is anticipated to be the first large-scale renewable technology to be able to deploy without financial support at some point in the mid-to-late 2020s.'

That had implications for policy, as also did the fact that, as DECC said, 'solar PV is a technology which can be deployed quickly even at large scale' (DECC 2014b). As we shall see (in Sect. 6.3), and as this review also noted, that had become a problem, with the consumer pass-through cost of its rapid expansion under the FiT being considered as too high. So FiT cuts followed. Nevertheless, *technologically* PV was now seen as a front runner, with Energy Minister Greg Barker clearly an enthusiast.

Wind power had featured strongly in the 2011 DECC roadmap, offshore especially. However, despite being the cheapest of the main new renewables, there were pressures on the new government to slow the deployment of onshore wind. In the 2014 review, looking at what might be relevant for the upcoming support programme, DECC provided a rationale for cutting support for onshore wind: 'Current installed capacity in the UK is 7.3 GW, with a further 1.5 GW under construction. There is also a large potential pipeline of UK projects with 5.41 GW having received planning consent and a further 6.5 GW currently in the planning system. This means we are well on our way to reaching our ambition for 11–13 GW of onshore wind by 2020.' So there was no need for new projects beyond what had already been supported.

By contrast, in its 2014 future support review, DECC saw offshore wind as a major potential growth area: 'Offshore wind is the most scalable of the renewable technologies, and it is the renewable technology that has the most potential to make a significant contribution to decarbonisation goals, if required. There is significant long-term potential for cost reduction and it is at an early stage of deployment – DECC's central estimate is a 25–30% reduction in central costs by 2030, which could be higher depending on the level of deployment between now and then. The UK is the market leader for offshore wind, with the biggest pipeline to 2020, and deployment in the UK is therefore a key driver of cost reduction to 2020.' DECC said up to 39 GW was possible by 2030.

Geothermal was not mentioned in DECC's 2011 'key options' list, but a 2012 SKM consultants study had claimed that it could in theory supply 20% of UK electricity from around 9.5 GW of capacity. The 2014 DECC support review however relied on a 2013 Atkins report on deep geothermal power, which only suggested a possible best-case potential of up to 3–4% of current average UK electricity demand. So it was marginal. Although some of the established biomass options (landfill and sewage gas, energy from waste combustion) were considered to have limited potential for expansion, they, along with advanced gasification/pyrolysis, biomass CHP and AD biogas from farm and other wastes, were all seen as eligible for some more future support.

DECC's 2011 Renewable Energy Roadmap suggested that there could be up to 300 MW of marine capacity (wave and tidal stream) by 2020. That was much less than the 1–2 GW forecast in the Government's Marine Energy Action Plan 2010, and, although the UK was still in the lead in this area, the 2014 DECC support review reduced its expectations further: 'Wave and tidal stream technologies are still at the demonstration stage and are not currently competing in the mainstream market. There are currently around 10 MW of wave and tidal stream capacity deployed in sea trial around the UK – more than the rest of the world combined. We anticipate that by 2020, wave and tidal stream could reach 100–150 MW in the UK alone. This deployment could then increase quickly beyond 2020 to reach GW-levels in the late 2020s–early 2030s.' So they remained long-term options.

For the sake of completeness, mention should also be made of yet another (and more positive) report on marine renewables, which came out between the 2011 DECC 'Roadmap' and 2014 support review. The Energy and Climate Change committee's 2012 report, 'The Future of Marine Renewables in the UK', said that, although costs were still very high, that should be seen as a challenge, and the government should increase support for wave and tidal power to preserve the UK's global leadership. The Committee had been told by industry representatives that the 2011 DECC Roadmap was 'too cautious' and 'pessimistic' and that 'a perceived downgrading of ambition from Government could have a negative impact on market confidence about the long-term future of wave and tidal energy in the UK' (Energy and Climate Committee 2012). As can be seen, DECC's 2014 support review, although not dismissive, arguably did little to change that, pushing their possible lift-off to the late 2020s/early 2030s.

It is also worth noting at this point that the long-running debate over the Severn tidal barrage finally came to an end. The Severn Barrage had received conditional support in a 2007 review by the Sustainable Development Commission, and some subsequently had hoped that it might get support under the RO or some form of hybrid private-public finance. However, in 2010, following a review looking at a range of tidal options, the new Conservative-led coalition government decided against providing support. It concluded that it did 'not see a strategic case to bring forward a tidal energy scheme in the Severn estuary at this time', although it wanted 'to keep the option open for future consideration'. It said, 'the decision has been taken in the context of wider climate and energy goals, including consideration of the relative costs, benefits and impacts of a Severn tidal power scheme, as compared to other options for generating low carbon electricity'. It noted that 'the Severn Tidal Power feasibility study showed that a tidal power scheme in the Estuary could cost in excess of £30 billion, making it high cost and high risk in comparison to other ways of generating electricity' (DECC 2010b).

There was a subsequent attempt by Hafren, a private developer, to push for a revised barrage plan, but nothing came of that, with the government (and a Select Committee) being unconvinced by the economics. Interests remained in smaller less costly barrages on other estuaries (in 2011, Peel Energy and the North West Regional Development Agency carried out a feasibility study for a barrage on the Mersey), and also in tidal lagoons, but, for the moment, the tidal barrage/tidal range option seemed to be stalled. With the decks cleared of that bit of history, the emphasis was now to be on finding a way to support more economically viable renewables, as well as other low-carbon options.

6.3 The CfD: And the Demise of the FiT

Although, as Box 6.4 describes, some specific technology priorities were shifting, the main focus of the Conservative-Lib Dem coalition government, which replaced Labour in 2010, was on the wider policy and support framework. It produced a 'National Policy Statement on Energy' in 2011, which said that, by 2025, the UK would need 113 GW of electricity-generating capacity, of which at least 59 GW would have to be new capacity, with renewables at around 33 GW, but 26 GW being 'for industry to determine', although it also said that it believed that 'in principle, new nuclear power should be free to contribute as much as possible towards meeting the need for around 18 GW of new non-renewable capacity by 2025' (NPS 2011).

The result was that nuclear, renewables and also CCS were all to be supported in parallel. To that end, as part of the government's series of Electricity Market Reforms (DECC 2012b), it proposed to set up a new support system to replace the RO. The new 'Contracts for Difference' (CfD) support system would cover nuclear, CCS and, from 2017, renewables (DECC 2014c).

This new contract system involved setting 'strike prices' for projects, with an agreement struck between power sellers (the generation companies) and power purchasers (the power supply companies), requiring the seller to reimburse the buyer if the actual market price for power at any point was less than the strike price, and the buyer to reimburse the seller if the market price was higher. As we shall see, after an initial trial CfD run for renewables without constraints, the government added a competitive contract auction element and also set cash limits to the overall CfD process, so as to protect consumers from high cost pass-through. Consumers would however still be faced with the extra costs for non-fossil energy (nuclear or renewables), or power from fossil projects with CCS. As with the NFFO, RO and the FiT, these costs would be passed on to them via their power bills.

The renewable CfD allocations were to be split between the developed and the less-developed options, with different initial reference (indicative) strike prices set. The technologies in the so-called Pot 1 established technology category were onshore wind (>5 MW), solar PV (>5 MW), energy from waste with CHP, hydro (>5 MW and <50 MW), landfill gas and sewage gas. Those in the Pot 2 'less established' technologies were offshore wind, wave, tidal stream, advanced conversion (biomass) technologies, anaerobic digestion, dedicated biomass with CHP and geothermal. DECC said that, for established technologies, 'the proposed move to competition builds on the strong progress on cost reduction and the well-developed pipeline. For less established technologies, our aim is to ensure that that they can deploy at levels which enable continued cost reduction and which ultimately support cheaper deployment in the long-term.'

DECC decided that it would not make wave and tidal stream technologies compete with other technologies for CfD support contracts, and would instead set aside a CfD funding slot (worth around £11 million) for at least 100 MW of marine power: 'Technologies such as wave and tidal stream are still at the demonstration stage and are not currently competing in the mainstream market. Introducing competition at this stage could mean that the industry is not able to develop at all' (DECC 2014d).

As can be seen, quite apart from added-on exceptions like this, the CfD was a quite complex system, which not everyone agreed would work well. At the overall level, was it sensible to have one system trying to support such widely different options as nuclear, CCS and a range of renewables at different stages of development? More specifically, while the CfD system attempted to reduce investment risks for generators, this might not work too well for those using variable renewables and might actually end up costing more than an FiT (Newbery 2011).

Muddying the water a little, even though, when market prices change, consumers would still face variable CfD reimbursement charges, the government confusingly sometimes labelled the CfD as an FiT system, presumably on the grounds that, once contracted, the strike prices for power fed in were fixed. But it was not an Obligation-based system like the

RO. No capacity targets would be set, that was left to the market, an approach the government saw as the next logical step.

Even so, given the complexity, it would take time to deploy fully, with the RO continuing meantime. So, while the transition to the CfD system was being prepared, changes were made to the RO, with DECC reducing support for onshore wind by 10%, on the grounds that it was getting cheaper. The existing FiT system for small renewable energy projects was retained, but it was kept as marginal, and, as we shall see, later cut back. Instead, CfDs were to be the main focus in terms of renewable support for new projects, with the RO to be closed to new projects from 2017.

In April 2014, in a preliminary CfD test round, under the so-called Final Investment Decision (FID) *Enabling for Renewables* process, 8 projects, 4.5 GW in all, including 5 offshore wind farms, were given contracts based on pre-set strike prices, with DECC providing outline price evolution estimates as shown in Table 6.4. The Treasury said 'these prices are broadly comparable to the support levels available under the Renewables Obligation, but include adjustments to account for the increased certainty generators have from not bearing wholesale price risk', this, it claimed, saving an expected £5 billion over the 15-year CfD operating period to 2030 (Treasury 2013).

This first trial run was certainly a large allocation, helping to draw down around £16 billion worth of investment, sufficient, DECC said, to help meet 14% of the UK's 2020 renewable energy target. It was seen as

Table 6.4 Draft strike CfD prices for renewable technologies

(£/MWh 2012 prices)	2014–2015	2015–2016	2016–2017	2017–2018	2018–2019
Onshore wind	100	100	100	95	95
Offshore wind	155	155	150	140	135
Tidal & Wave	305	305	305	305	305
Biomass conversion	105	105	105	105	105
Solar large	125	125	120	115	110

Treasury (2013)

They were later revised slightly, with onshore wind prices falling to £95/MWh up to 2016–2017 then £90 afterwards. Large solar fell to £120/MWh in 2014–2015 and then in stages to £115 and £110 and finally £100 by 2018–2019 (DECC 2014e)

Box 6.5 The First Full Renewables CfD Round

The first full round with competitive capacity auctions, run in 2015, led to £315 million in contracts being awarded for 39 projects. Two offshore wind farms in Scotland, a total of 1162 MW capacity, got strike prices of £114.39 and 119.89/MWh. Fifteen onshore wind farms got contracts for a total of 748.55 MW capacity, with average strike prices varying between £79.23/MWh and £82.50/MWh. Two Energy from Waste/CHP projects also got contracts (95 MW total), along with 3 advanced biomass conversion projects (62 MW in all), and 5 solar PV projects (72 MW in all), with very low strike prices: two at £50/MWh, almost at convention power cost, the other three at £79/MWh, still lower than the £120 DECC initially specified in its Reference price levels. Overall, DECC said it had resulted in the capacity delivered costing up to £105 million less per year than it would have in the absence of competition (DECC 2015a).

having overcome what was felt to be an 'investment hiatus', and moving things on, while the full CfD process was being set up. However, with this early allocation having helped establish the market, and with concerns about the likely cost to consumers also rising, it was decided to introduce a competitive auction process for the first full CfD round, in 2015. Some quite low strike prices then resulted. See Box 6.5.

Clearly, competition was pushing prices for renewables down. That had always been the aim. In 2011, the government had said that it was 'minded to move from administrative price discovery processes for low-carbon technologies to more competitive forms of price discovery such as auctions or tenders when the wider conditions in the market will support their successful deployment' (DECC 2011b).

Controversially, that evidently did not apply to nuclear, with the 3.2 GW Hinkley project being given a CfD contract in 2013 without going through a competitive tendering process. The nuclear industry had professed not to need subsidies, but it did want the market structure to be supportive, and some saw the Electricity Market Reforms, and the CfD, in that light. DECC had claimed that, overall, renewables would be given similar treatment to nuclear, and that the nuclear CfD confirmed broadly with the government policy of offering no public subsidy to nuclear, even if it might differ from some of the other contracts agreed.

However, the status of that promise began to look a little uncertain, given that Hinkley seemed to be given very privileged status. The CfD strike price offered, with no tendering required, to French company EdF for Hinkley, was £92.5/MWh, inflation index linked, with guaranteed support for 35 years, as opposed to 15 years for CfD contracts for renewables (HM Government 2013; HM Treasury 2013). Moreover, the CfD support level for Hinkley was higher than for the new onshore wind and PV projects, which would start up soon, and also seemed to be higher than what was likely to be needed by offshore wind by the time the Hinkley plant might start up (currently put at 2027), if it did go ahead. In addition, as a final extra level of support, a £10 billion loan guarantee was offered to EdF for the Hinkley project, under the UK's infrastructure support system (HM Government 2012).

Overall, it did seem that the approach to nuclear/renewables support, at least at that stage, was asymmetrical (no CCS projects have as yet gone ahead), and initially it seemed likely that the funding arrangements for the Hinkley project would fall foul of European Commission's state aid rules. However, after some adjustments, with some profit claw-back arrangements added, the EC gave it a go-ahead (Europa 2014).

So all was set for the UK government to press ahead with nuclear. Several other nuclear projects were proposed, from EdF and other overseas companies, around 16 GW in all, with some expected construction start dates in the late 2020s/early 2030s. By contrast, on the basis of data in a 2013 DECC report, after expansion up to 2020, so as to meet the UK's 15% renewable energy target (as agreed with the EU), the growth of renewables would fall off, with less new capacity being added (DECC 2013b). See Fig. 6.2 for a rendition of this. Moreover, with BREXIT likely well before then, even the 2020 15% target might not be honoured, and, with cash caps also likely to be tightened, the future might not be too good for renewables.

Gloomy prognoses like that were enhanced by a nuclear strategy review from the newly established (and short lived) Department for Business, Innovation and Skills (BIS), which suggested that 75 GW of nuclear might be possible by 2050, supplying 86% of UK electricity, up from the then around 20% (BIS 2013). BIS ran in parallel with DECC for a while,

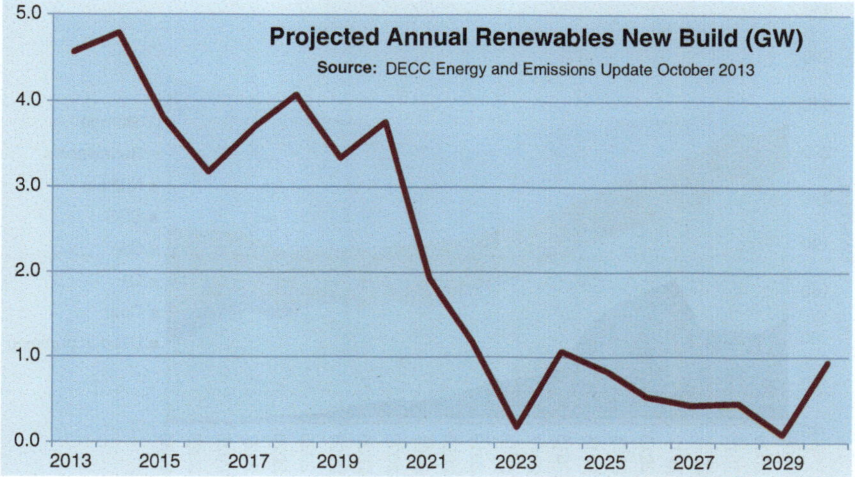

Fig. 6.2 Capacity growth fall—pessimistic view. (Data from DECC 2013b spreadsheet)

but following the election of a Conservative Government ruling alone from 2015, in 2016, both BIS and DECC were abolished and replaced by a Department for Business, Energy and Industrial Strategy (BEIS). DECC had been a Liberal Democrat stronghold, and there were concerns that, with the strongly pro-renewables Lib Dems no longer sharing power, renewables would take a back seat, with, in effect, the rather grim Fig. 6.2 prognosis coming true. However, that was perhaps not inevitable. The main scenario offered in DECC's 2013 report (Figs. 6.3 and 6.4) was much more optimistic: renewables could expand rapidly. So which way would it go?

In the event, the nuclear push seemed to be blunted to some extent by a combination of problems with nuclear and success with renewables. EdF had major problems with the EPR plants it was building in France and Finland (the same design as that proposed for Hinkley) with large cost overruns and long delays and, around the world, nuclear developers faced financial problems, so that securing investment for the UK's proposed follow-up plants proved to be difficult.

By contrast, the renewables seemed to be going from strength to strength, winning contracts under the CfD auction system at increasingly

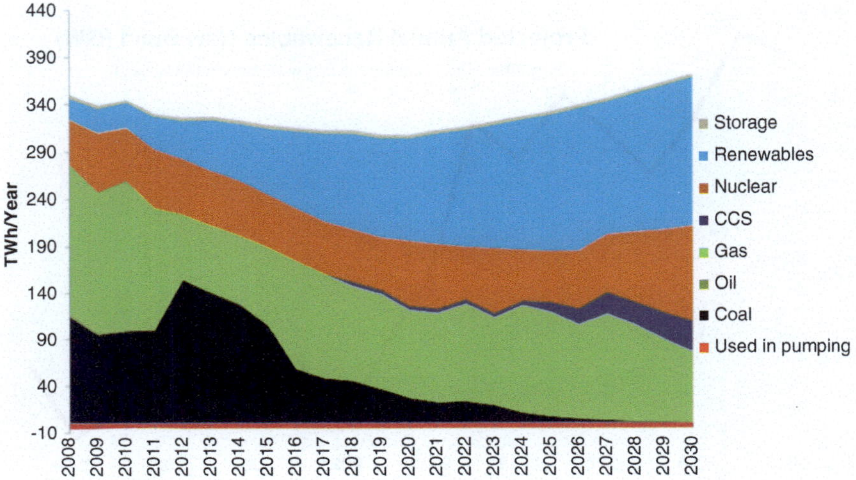

Fig. 6.3 DECC's electricity supply projections. (DECC 2013b)

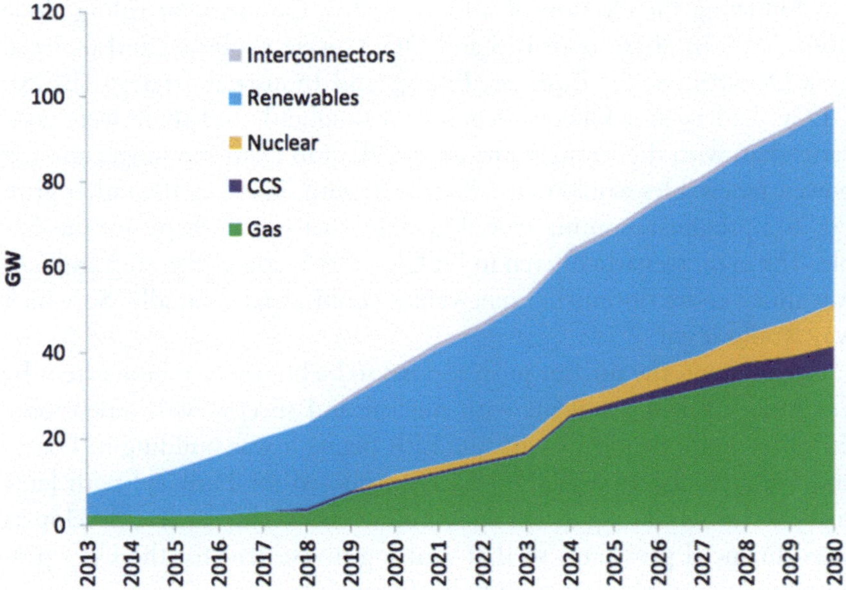

Fig. 6.4 DECC's new-build capacity projections. (DECC 2013b)

low strike price levels, as for offshore wind in the second CfD round, in 2017, discussed below, with capacity building up. So much so, that a BEIS scenario running up to 2035, produced in 2018, had renewables expanding quite rapidly, eventually to 68 GW in their reference scenario, while fossil CCS, at one time the third major element in the programme, was almost eliminated (BEIS 2018b). That no doubt reflected the decision in 2015 to abandon the UK's £1 billion CCS programme and similar moves elsewhere in the world (Elliott 2018). So CCS was out of the running for now, leaving nuclear and renewables as the main players in the UK. See Fig. 6.5 for a version of the BEIS scenario as rendered by Carbon Brief (Evans 2018). Note the rapid phase out of coal (even more rapid than in Fig. 6.3) and also the significant decline of gas. But also note that the rapid expansion of renewables was not quite as much as was envisaged in DECC's 2013 more optimistic scenario (Fig. 6.3), with nuclear doing somewhat better.

Even so, a new balance seems to have emerged. Certainly CCS was out. The then Prime Minister David Cameron said 'You spend £1 billion on carbon capture and storage, you get some carbon capture and storage capacity and it would cost you, at the current estimate, something like £170 per megawatt-hour. That compares with unabated gas costing £65, onshore wind perhaps as costing £70 and nuclear costing, say, £90. […] Governing is about making decisions, and it seemed to me that the right decision was to say that we would not go ahead with the £1 billion, because that is £1 billion that we can spend on other capital investment projects, including energy projects such as making progress on energy storage or modular reactors' (Cameron 2016).

As can be seen, he was leaving the door open to new nuclear, including the longer-term possible option of small modular reactors (SMRs). But his portrayal of nuclear as more expensive than onshore wind is interesting. That clearly was what emerged from the CfDs, with offshore wind and PV CfD strike prices falling. That did not mean that nuclear was out of the picture, as the BEIS chart for individual technology growth rates shows (Fig. 6.6), but renewables were seen as moving well ahead and output growing, although with nuclear accelerating too, as and when the proposed new projects got going.

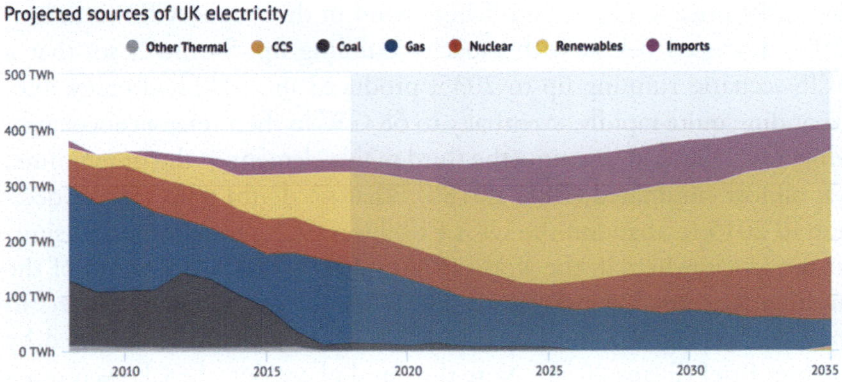

Fig. 6.5 Version of BEIS scenario. (Evans 2018, based on BEIS 2018b)

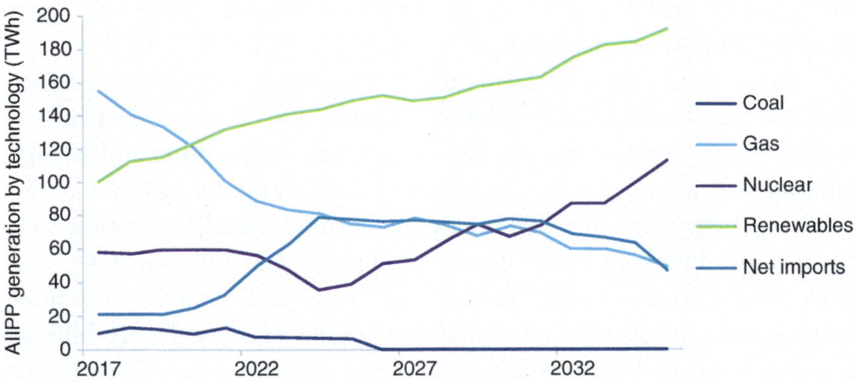

Fig. 6.6 BEIS energy mix projections. (BEIS 2018b)

That might seem a little speculative, given the uncertainty over finding funding for them, and their costs. By contrast, the renewables seemed much better place to expand, with costs clearly falling. Cost reductions had certainly been slowly becoming apparent, but when they become clear from the continuing CfD contract successes, they took some observers by surprise. It had been said so many times that renewables were expensive. See Box 6.6.

Box 6.6 Power Cost

One of the more common ways to render energy costs is via an estimate of the lifetime cost of generation from a plant. The cost of the plant, fuel and running costs are averaged out over its working life. This so-called Levelised Cost of Energy (LCOE) approach has problems. For example, it is often hard to predict future fuel costs, much less changing interest rates on any capital borrowed to build the plant. It is also hard to make comparisons between different technologies on this basis. They may have very different capital cost structures, and, in the case of renewables like wind and solar, no fuel costs. Nevertheless, comparative LCOE data is widely used, as in the set below, used by the UK government, initially produced by consultants Mott MacDonalds. See Table 6.5.

Table 6.5 Generation cost estimates per MWh delivered, in the UK based on Mott MacDonald data, levelised costs in £/MWh

Electricity option	Current cost	Cost in 2040
Onshore wind	83–90	52–55
Offshore wind	169	69–82
Tidal barrage	518	271–312
Tidal stream	293	100–140
Wave (fixed-floating)	368–600	115–300
Hydro (small/run of river)	69	52–58
Solar photovoltaics (PV)	343–378	60–90
Biomass (wastes/short rotation coppice)	100–171	100–150
Biogas (anaerobic digestion/wastes)	51–73	45–70
Geothermal	159	80
Nuclear (water cooled reactors)	96–98	51–66
Gas with carbon capture and storage	100–105	100–105
Coal with carbon capture and storage	145–158	130

Mott MacDonald (2011)

The data in Table 6.5 is not definitive: critics suggested that the estimates for nuclear were very optimistic (being for as yet unbuilt plants of new types) and those for renewables pessimistic, given price reduction trends for actual projects. Certainly Mott Macdonald said they were 'bullish' on nuclear costs. And since 2011, PV costs have fallen dramatically to be more like the 2040 estimates in the above. Similarly for offshore wind.

LCOEs continue to be used, for example, for projections and comparisons, and may have some advantages over the so-called overnight cost approach, which assumes plants can be built instantly, overnight, so that no account is taken of the real-world cost of borrowing money and of dealing with future cost changes.

Note that CfD strike prices are clearly not the same as LCOEs or overnight costs, but they do have the virtue of representing what buyers and sellers think is a reasonable price to set. However, neither they nor LCOEs or overnight costs cover all the system and grid balancing costs and, in some cases, these can be significant. We will be returning to this issue in Chap. 7.

However, the new CfD results meant that the view of renewables as 'high cost' was no longer tenable. In the second full CfD round in 2017, two offshore wind projects with 2022–2023 start-up dates were offered strike prices of £57.5/MWh, a 50% fall from the previous CfD round in 2015. A third offshore wind project in the new round was offered £74.5/MWh, making the offshore wind contract total 3.2 GW (BEIS 2017b). Price reductions like this had in fact already occurred elsewhere. For example, a Danish offshore wind farm had got a supply contract at under £44 per MWh, or around £55 per MWh including grid connection costs. But the CfD price falls still came as surprise. The Times editorial (12/9/17) said 'the contrast with Hinkley Point could hardly be any more stark'. Since the £95.5/MWh Hinkley strike price was index linked, with inflation, it would be more like £100/MWh, and rising, by the time it might be running perhaps in 2027. By that time, further CfD rounds could presumably have delivered offshore wind projects at even lower than half the eventual strike price that Hinkley would enjoy.

The comparison was even more striking when looking at onshore wind, which would likely be even cheaper. In the first full CfD round in 2015, onshore wind strike prices had ranged from £79.23 to 82.50/MWh (DECC 2015a), and the technology had improved since then.

In 2017, there were projects going ahead in Germany at €38/MWh average, with lowest bid in its third onshore wind contract auction (for 1 GW of capacity) being at €22/MWh (Business Green 2017).

While the prospects thus looked good, the problem in the UK was that there was pressure from opponents of wind farm deployment and wind subsidies, including a letter to the Prime Minister from over 100 MPs, all but 5 being Conservatives (Telegraph 2012). In response, the government introduced new tougher planning policies, making it harder for onshore wind projects to go ahead, with the planning minister also stepping in to halt some projects. For example, Communities and Local Government secretary Eric Pickles reportedly turned down 19 onshore wind farm applications (Merrill 2014).

In addition, in its Election Manifesto in April 2015, the Conservative Party made a clear commitment to end public subsidy for onshore wind farms and to 'halt the spread of on-shore wind', if elected. Subsequently, following its return to power without the Liberal Democrats in May 2015, the new Conservative Government blocked new onshore wind projects from access to the RO, and also added more planning constraints, making it easier for objectors to block onshore wind projects and for the planning authorities to take action. A new even tougher planning environment was thus created.

Even though 'grace' periods were agreed for projects already nearing RO contracts, the new policies were a major shock for wind developers, with planning applications for new onshore wind developments plummeting by 94%. The Energy Secretary Amber Rudd had talked of 7.1 GW being blocked, about 2500 turbines (Rudd 2015), and it might end up being more than that, with investor confidence falling.

However, worse was to come. With access to the RO blocked, the CfD was now its main route forward, and, as we have seen, onshore wind projects did well in Round 1, but it was decided that the second CfD round, in 2017, would be focused just on Pot 2 technologies, so the government also effectively blocked onshore wind from further access to the CfD. Clearly, falling costs were part of the reason, but the government was also faced with escalating consumer pass-through costs from the CfD. I will be exploring the government's green levy cash-limits issue later (it imposed a cap on spending), but, from the wind industry side, the decision to block onshore wind from access to the CfD was a further major blow. In response, it argued that CfD contracts, even if with zero subsidies, were vital to keep the technology in the market, but the policy stayed, as did the planning constraints. Those limits, and the RO and CfD blocks, would of course have knock-on effects. For example, the Citizens Advice Bureau said the block to onshore wind might cost consumers £0.5 billion, since more expensive sources would have to be used to meet the 2020 renewable energy target (CAB 2017). However, there was no going back, onshore wind was blocked.

Wind farm developers were not the only ones to face new policies. A similar approach was adopted for PV solar. That too was now becoming cost competitive, especially in the case of larger solar schemes, as was indicated in the first full CfD round in 2015 (see Box 6.2), in which two large PV projects got very low strike prices, £50/MWh, almost at conventional power cost, and three others came in at £79/MWh.

The acceptance of these projects for CfDs turned out to be exceptions; they had slipped through a closing net. In 2014, the coalition government had put planning blocks on some solar farm schemes, on visual intrusion and land-use grounds. Energy Minister Greg Barker said 'I do not want solar farms to become the new onshore wind. I do not want to see unrestricted growth of solar farms in the British countryside', adding that 'we don't want the whole solar sector damaged by a few solar farms that communities don't want' (Barker 2014). Planning constraints were therefore imposed. See Box 6.7.

Box 6.7 Solar Farm Planning Blocks

In 2014, the Department for Environment, Food and Rural Affairs (DEFRA) blocked farmers' access to subsidies for fields if they had solar panels from 2015 onwards, so as to 'ensure more agricultural land is dedicated to growing crops for food'. Farmers using fields for PV arrays could no longer bid for payments for that land under the EU Common Agricultural Policy. DEFRA said this would save 'up to £2m of taxpayers' money each year that won't be available for these subsidies'. It added 'The changes the government is making are expected to slow down the growth of solar farms in the countryside in England. There are currently 250 installed, with the biggest covering as much as 100 hectares. Under previous plans, the number of fields dedicated to solar farms was set to increase rapidly, with over 1,000 ground-based solar farms expected by the end of the decade across the UK. These changes should help to halt this expansion as it will now become less financially attractive for farmers to install the solar panels' (DEFRA 2014).

The Environment Secretary, Liz Truss, said 'Solar panels are best placed on the 250,000 hectares of south facing commercial rooftops where they will not compromise the success of our agricultural industry', noting, in an interview with the *Sunday Mail*, that it was 'a big problem if we are using land that can be used to grow crops, fruit and vegetables' (Walters and Carlin 2014). However, when asked in Parliament what estimate DEFRA had made of the area of land occupied by solar arrays that was arable land useable for economically farmed fruits or vegetables, a DEFRA minister admitted 'we have made no estimate of the amount of land occupied by solar panels which was arable land useable for economically farmed fruits or vegetables' (Murray 2014).

There had certainly been some local complaints about visual intrusion, but there were also dissenting views about the likely environmental impacts of solar farms in rural areas. It was argued that they could still allow for sheep grazing and create wild flower and wild life havens (Montag et al. 2016).

The government's opposition to solar farms was not just on visual intrusion and land-use grounds. After all, there were already planning controls and processes that should be able to avoid inappropriate siting. It was also related to strategic economic issues. A block on solar farms would save CAP money, and avoid passing too much cost on to consumers. So, in addition to the planning constraints, in 2014, the RO was also closed for new PV capacity above 5 MW from April 2015, although with a grace period to protect developers who had already made financial commitments. DECC said that closing the RO to solar farms would save up to £100 million a year from the RO budget by 2017 and more later. Otherwise, they might reach 6.3 GW by 2020, and so add substantially to the pass-through RO costs to consumers and break DECC's budget allocation for renewables support (Shankleman and Murray 2014).

The solar industry was dismayed by these moves. One of the leading solar installers, Solar Century, said 'Large-scale solar is already significantly cheaper than offshore wind and will be competitive with onshore wind by 2017. In deliberately setting out to strangle the growth of cheaper solar from 2015, Secretary of State Davey can no longer claim that government policy will deliver the most cost-effective mix of technologies by 2020' (Solar Century 2014).

As a concession, DECC had said that large PV schemes could still apply for CfD support, but, subsequently, after the new Conservative Government took power, the position changed. As noted earlier, it was decided to focus the 2017 CfD round just on Pot 2 technologies, so large PV projects, as well as onshore wind projects, were in effect blocked from access. Although this decision no doubt reflected the view that large solar farms were inappropriate, there were also economic grounds: the rapid expansion would cost too much, overwhelming the budget allocation, an issue we will explore more later. Overall then, for whatever reasons, these moves amounted to almost a full halt to solar farms and large solar arrays, apart from projects that were deemed appropriate, for example, on factory roof tops, and even those would not get CfDs.

While this policy may have had some environmental justification, it was harder to see that in the case of the parallel moves in relation to small-scale PV projects. They were not eligible for CfDs or the RO, but did get support from the FiT system. However, the government decided it should be cut back. The full story is somewhat convoluted and spread over a number of years. The FiT was launched in early 2010. Subsequently, domestic project take-up boomed much faster than expected, overwhelming the pre-set price degression rates. That would push the pass-through cost to consumers up even more, and an FiT cut-back was therefore felt to be needed, especially since, with costs falling, new projects should no longer need so much support. So, in October 2011, Energy Minister Greg Barker said: 'We must reduce the level of FITs for solar panels as quickly as possible, to protect consumer bills and to avoid bust in the whole Feed-in Tariff budget', and he proposed to do just that (DECC 2011c).

However, the scale of the FiT cuts that were subsequently proposed was provocative (the headline figure being a fall from 43.3 p/kWh to 21 p/kWh for retrofit PV), with PV developers seeing that as likely to slow down PV expansion drastically. There was a long drawn out battle, including recourse to the courts by Friends of the Earth, Solar Century and HomeSun. They won backing from a High Court Judge. DECC submitted an appeal, but it was denied, with the whole affair dragging on, acrimoniously. DECC talked of going to the Supreme Court, but that could lead to a substantial delay. There were further rounds of consultation, but in 2012, even more cuts were proposed (from 21 p/kWh to 16 p/kWh), although there was a small sweetener: the export tariff, at that point 3.2 p/kWh, would be increased to 4.5 p/kWh (Harvey 2012).

You may recall, from Box 6.4, DECC's 2014 view that PV solar would soon be able to go forward with less subsidy and 'without financial support at some point in the mid-to-late 2020s'. That was now to be put to the test, but at a very much earlier stage. With a Conservative administration in power from May 2015, and concerns about green levy budget overspend increasing, although minor concessions and a delay were agreed, the basic policy remained unchanged and indeed was accelerated and toughened up dramatically. There would be very major cut to the FiT, and then the whole FIT system would be abolished in 2019. See Box 6.8.

Box 6.8 The FiT Cuts—PV Gets Hit Hard and Is Then Shut

With the pass-through cost to consumers rising, while PV cost themselves were falling, the government's approach was quite aggressive, with at one time an 87% cut being proposed. After widespread protests, this was reduced 64%, but worse was to come, with this being only a temporary respite. The Government said 'the scheme will remain under review to ensure it continues to achieve its objectives until generation tariffs end in 2019' (DECC 2015b).

In its consultation on the longer-term future of the FiT, DECC had certainly taken a hard line, possibly reflecting the change in Government. It said that 'if such measures cannot put the scheme on an affordable and sustainable footing then there should be an end to generation tariffs for new applicants as soon as legislatively possible, which we would expect to be January 2016'. But even if all went as they wanted, they saw their proposed 'more stringent degression mechanism and deployment caps' as 'leading to the phased closure of the scheme in 2018–19' (DECC 2015c). Not much of a consultation: take it or leave it!

Thus, DECC said 'If following the consultation we consider that the scheme is unaffordable in light of these criteria, we propose ending generation tariffs for new applicants from January 2016 or, alternatively, further reducing the size of the scheme's remaining budget available for the cap. This consultation seeks views on the impacts of scheme closure, whether implemented in the immediate term or as a phased closure over several years.'

Following the consultation, it was decided that the scheme could continue beyond 2016, albeit with much lower tariff rates, the best PV price (for under 10 kW) being 4.39 p/kWh, moved up from the earlier proposal of just 1.63 p/kWh. However, that concession, and lesser ones for smaller PV projects and small wind, was just for now, with the 2019 complete FiT closure plan still going ahead.

The end to the FiT was confirmed in 2018, with BEIS now having replaced DECC. No immediate replacement was proposed. BEIS said it felt that that 'consumers should not be expected to support sectors indefinitely and that as costs fall, so too should support'. This also included the export tariff, which was much smaller (at least than the initial production tariff) and so less provocative, and might be seen as still very useful, feeding power to the grid. However, BEIS said that 'we do not believe that the current FIT flat rate export tariff aligns with government's vision for the future, given our desire to move towards fairer, cost reflective pricing and the continued drive to minimise support costs on consumers' (BEIS 2018c).

In theory, with PV costs now lower, new projects could still come through, but unsurprisingly, the cuts led the number of PV planning applications to fall. According to DECC data, there was a 78% drop in applications lodged in June 2015 compared to the previous year, and that trend showed up in the eventual level of deployment: see Fig.6.7. The slowdown has continued since.

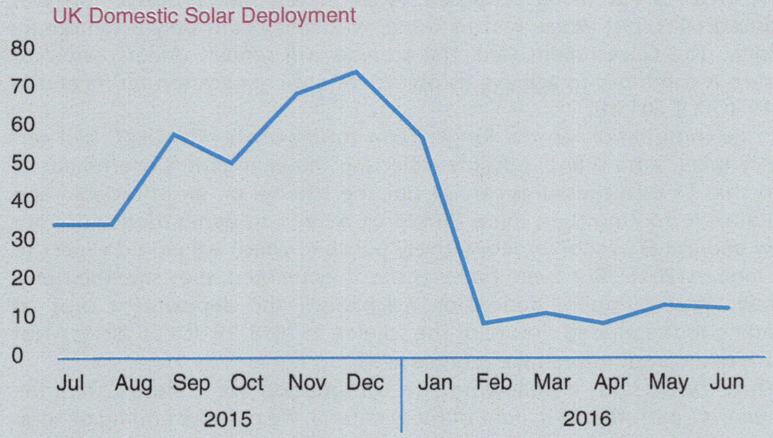

Fig. 6.7 Domestic PV MWs installed each month, based on BEIS data. (BEIS 2017c)

PV solar was not the only target for FiT cuts. PV was the largest user of the FiT system, but cuts were also sought for small wind and hydro projects, leading to yet more conflicts and contestation. Subsequently, and perhaps even more contentiously, it was proposed to cut the FiT for generation using biogas by 27% for small to medium power plants and *entirely* for larger (500 MW and above) plants. It was major shock for the fledgling biogas sector (Whitlock 2016). Given the strong reaction, the cuts were partly fought off (BEIS 2017d), but of course the whole FiT system would end in 2019, so, unless policies changed, the sector would have to refocus, possibly to biogas for heating and transport use: see Sect. 7.6.

As a background to this rather sad saga of cuts, it had earlier been announced (in the 2015 Budget) that renewable electricity would no longer be exempt from the CCL on business energy use. Levy Exemption Certificates (LECs) generated as a result of the CCL, which was originally designed to promote the generation of clean energy, have provided vital financial support for renewable energy producers. The Chancellor said the removal of the exemption would earn the Treasury £450 million in 2015–2016, rising to £910 million in 2020–2021, and there were now other support routes for renewables, which did not, like the CCL, risk some (about a third) of the benefits going abroad. With this surprise move likely to cost the sector £3.9 billion over the next five years, Renewable UK were totally dismayed: 'We're suddenly looking at a substantial amount of lost income for clean energy companies which was totally unexpected. For example, Levy Exemption Certificates account for just over 6% of onshore wind generators' revenues. It's another example of this Government's unfair, illogical and obsessive attacks on renewables' (Tickell 2015).

Certainly the combined effect of this and all the blocks and cuts for PV and onshore wind was to call into question whether the optimistic scenario for renewables expansion that BEIS had produced was really viable. Could it all be done by offshore wind and some large biomass conversion plants? Offshore wind was doing well, with some impressively large projects being installed around the UK coast, in all around 5 GW by 2016, with more planned and machines of 8 MW and more being installed. New ideas were also being developed for floating offshore wind schemes, which would avoid the cost of building foundations and allow projects to go further out to sea, where the wind regime was usually more attractive. However, the other main option, biomass conversion, was quite controversial, since some of the large plants relied on using wood pellets imported from North America, with critics alleging that the approach being used was not sustainable (NRDC 2016). There were strong environmental objections to the use of what was seen as an important carbon sink (Brack 2017). Nevertheless, the government seemed wedded to this option, as long as it was subject to proper environmental controls. Certainly, it delivered a large amount of energy from old coal-fired plants converted to using wood as a fuel, and the biomass trade lobby remained quite strong (REA 2017).

As can be seen, the CfD programme, and the selection of projects eligible for it, had become highly politicised, especially for wind and PV, and also biomass, that is, the three market leaders. More generally, even leaving the cuts aside, it was not clear whether the CfD system would be any more successful at stimulating the deployment of renewable capacity than the RO. That also applied to support for the newer, less-developed, renewable technologies. The first full Round of the CfD in 2015 did have a protected 100 MW slot for wave and tidal power, but no projects were able to get support. Much as with the RO, the CfD was evidently not suited to emergent technologies like this, even including tidal current turbines, which, by this stage, had become quite well developed. For example, Atlantis was in the process of installing a series of tidal turbines in Pentland Firth, as part of the planned (eventual) 398 MW Meygen project. The first set of 4 turbines, installed in 2017, had (as noted earlier) managed to get a contract under the RO (with 5 ROCs/MWh promised), but, with the RO scheme being phased out, it proved impossible to get support for next set of turbines under the more competitive CfD. It is unclear what will happen next.

It was the same for the proposed 320 MW tidal lagoon in Swansea Bay, although, in that case, the project would have needed a CfD at maybe £90/MWh running for perhaps 90 years, as opposed to the standard 15 years used for renewable CfDs and the 35 years agreed for the Hinkley nuclear plant. The 90-year figure was considered as reasonable by the developers, given the very long operating life of lagoons. Subsequent lagoon projects could be larger and in areas with a higher tidal range, and so could be cheaper, as was argued in a review produced for the government by one-time Conservative energy minister Charles Hendry, with the Swansea project depicted as a 'pathfinder' (Hendry 2016).

However, in the event, this first-of-a-kind scheme was seen as just too expensive, and it was blocked by Business and Energy Secretary Greg Clark, who argued that 'however novel and appealing the proposal is… the costs that would be incurred by consumers and taxpayers would be so much higher than alternative sources of low carbon power, that it would be irresponsible to enter into a contract' (BEIS 2018d).

Understandably, this decision was not well received by the devolved government of Wales, or by the tidal energy lobby, who thought it lacked vision (Early 2018), although it was supported by a study done for the National Infrastructure Commission by the Aurora consultancy, which concluded that 'tidal lagoon projects are unlikely to ever be economic without government support, unlike other renewables' (Aurora 2018).

Wave energy was an even more unlikely contender. Indeed, its prospects were looking poor. Although, in 2013, the Scottish government had offered £13 million to support for two projects, they did not go ahead, with the two leading developers, of respectively, the Pelamis and Oyster devices, both going bust. The Scottish government rescued some of the staff, setting up a residual programme, but the focus was on R&D and small system testing and development. A retrospective review of the wave energy programme criticised the way the wave energy programme had been handled (see Box 6.9), but offered a fairly bleak prognosis. That view was reinforced by a study of the prospect for marine renewables by the Energy Technologies Institute (ETI). It thought that tidal stream technology might prosper, but the approach to wave energy had to be rethought (ETI 2017).

Box 6.9 The Wave Energy Programme Reviewed Again

A 2017 review of the UK wave energy programme by Imperial College London and Strathclyde University claimed that there had been a premature emphasis on commercialisation. The requirement for state-supported developers to secure private-sector match-funding had brought 'intense investor pressure to reach commercialisation as quickly as possible' (Hannon et al. 2017).

The report says that a major institutional barrier was 'the overwhelming emphasis on full-scale device demonstration, with a view to 'fast tracking' progress to commercial array-scale projects before the underpinning early- to mid-stage R&D had been performed. Reasons for the UK going 'too big, too soon' included public and private sector funds being made available to progress the technology as quickly as possible following developers' highly optimistic claims about the promise of wave energy. The outcome was that developers had over-promised in order to receive funds but then subsequently under-delivered, in turn eroding investors' confidence in wave energy and reducing their willingness to invest in the technology, triggering the collapse of leading firms (e.g. Pelamis) and further undermining the sector's legitimacy. Underpinning these developments was a poor understanding of the scale of the innovation challenge and the associated time and funds required to overcome it, as well as a lack of rigorous, objective procedures to review the credibility of funding proposals.'

The study covered the period from 2000 to the present and concluded that the prospects did not look good. As yet, there was no sign of wave projects getting anywhere near the cost level that would enable them to compete for support under the CfD capacity auction system. The report saw at least some of this as being due to poor programme management: 'explicit government targets for wave energy deployment have steadily reduced in ambition before being removed altogether by the UK Government… Whilst there was a large number of UK government publications (e.g. white papers, parliamentary reports) calling for the need to support wave energy, there was a clear change in direction from the early 2010s, with a removal of formal wave energy deployment targets and a decline in vocal support from government ministers. This was at odds with the UK general public's support for wave & tidal energy, which averaged 74% since 2012, greater than the figure for onshore wind (67%) and equal to that for offshore wind (74%).'

Should new policies be tried? The UK wave energy resource was very large, and it seemed foolish to ignore it and the potential for exporting the technology to other countries with good wave resources. Although there was work going on elsewhere, the UK was still a leader in the field. Thus, *RenewableUK* said that this was no time to cut support: 'The UK is right at the forefront of a global race to develop wave power on a commercial scale. It's vital that we don't lose our lead to other countries, who stand to benefit from the years of investment and progress we've made. The more we deploy, the cheaper the technology will become' (RUK 2017).

That could be true, and, with this in mind, the report suggested that 'special case' treatment might be given to wave power work: 'Emerging technologies, such as wave energy, can be out- competed for subsidies on a cost basis when in direct competition with significantly more mature technologies. Specific examples include separating wave energy from the same […] CfD allocation as significantly cheaper technologies such as offshore wind energy and avoiding wave energy becoming bundled into wider marine energy RD&D programmes, where it must compete with more mature technologies such as tidal range and tidal stream.'

However, with the emphasis of UK energy policy being to go for market competition, it did not look very promising for wave energy. There were some devices under test, and there might be niche markets where some types of wave energy systems might prosper, and clever new ideas might continue to emerge, with, hopefully, costs coming down, as has happened with wind and PV solar. But for the moment, sadly, although some suggestions were made for improving its prospects, wave energy looked somewhat becalmed.

So, one way or another, there did not seem much chance of the CfD helping new options along. Moreover, for the rest, it was not clear if, given the policy framework in which it was set, the CfD would provide sufficient support to allow renewables to expand to meet the national and EU targets. In addition to blocks on large PV and onshore wind from the CfD, and the FiT cuts, as we shall see, the government also imposed tight cash limits on the overall programme. Was this really a viable system for overseeing expansion?

Given this situation, it is worth briefly backtracking at this point in the story to look at the underlying economic policy rationale for the cuts and cash limits, which was that, without them, the budget allocated under the governments 'Levy Control Framework' would be exceeded.

6.4 Funding Cuts and LCF Caps

As we have seen, after an initial trial run without cash limits, the operation of the CfD was set within a competitive framework. Overall spending caps were also imposed. They were designed to keep the CfD programme within the limits set by the annual 'Levy Control Framework' (LCF) finance cap, ostensibly to avoid high pass-through costs to consumers. While that was understandable, it did mean that the expansion of renewables would be constrained, unless the cap level was raised. It is worth exploring this issue in a little more detail, since it has shaped what came next.

The overall LCF limit was initially set for 2014–2015 at £3.5 billion, rising in stages to £7.6 billion in 2020–2021. That was for all so-called green levy schemes, including the RO, FiTs and the CfD. The problem was that the agreed spending looked certain to rise to around £9 billion by 2020–2021. See Fig. 6.8. Cuts would have to be made, and, as we have seen, they were.

From DECC's point of view, longer term, staying under the LCF cap would be made a little easier since, although contracted RO projects would be supported after 2017, when the RO scheme ended, no new projects would be supported after that. Similarly for the FiT, after 2019. Nevertheless, whatever happened next, pressure on the LCF would rise,

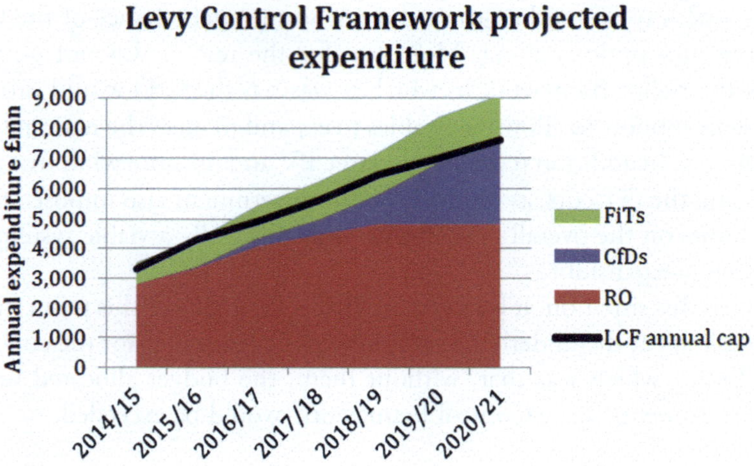

Fig. 6.8 LCF projections. (DECC 2015c)

since several CfD-backed projects were already in the pipeline. Their CfD contracts were agreed and could not be revoked. In particular, five offshore wind projects had got CfDs under the interim FID *Enabling for Renewables* trail round in 2014, before the caps were set. It was pointed out that this initial trial CfD run, without a competitive auction process, had led to much of the funding available (58%) for the next phase having been allocated (PAC 2014). DECC was nevertheless rather defensive about the trial run, which it said, perhaps a little lamely, 'ultimately enabled price competition to happen sooner than it would have otherwise been able to, through introducing and testing the CfD with the market' (DECC 2015d).

However, the overspend problem had been compounded by a perverse effect from the framing of the CfD system. Under the CfD, if actual market prices fall, suppliers would be given compensatory 'top ups', to the pre-agreed 'strike price', and this money would come from extra charges to consumers. That would show up as a rise in the LCF figures. This was one reason, the Department said, for the LCF overspend—market prices had fallen.

Clearly, unless action was taken, it might get even worse. In mid-2016, the National Audit Office noted that 'since 2012, the Department has revised downwards its projections of future wholesale electricity prices, mainly because of a global reduction in the prices of fossil-fuels. While CfDs reduce the risk to consumers of market price volatility, they also mean consumers benefit less from wholesale price falls, which are offset by increased top-up payments. The present value of future top-up payments through existing CfDs has increased by £5.6 billion in the 2015–16 financial year because of lower projected wholesale electricity prices' (NAO 2016).

To slow things down, and stay within the LCF limits, for the first full CfD round in 2015, the government imposed a spending cap of, initially, £205 million p.a. While some tried to argue against this, it was generally seen as necessary, although some called for the overall LCF cap level to be raised to allow for more CfD spending. However, in general, the cap was accepted as a *fait accompli*, although there were debates over how it should be applied *within* the CfD system. Who should get what?

The original cash allocation cap for Round 1 had been set up under two categories, running up to 2019. Firstly 'Pot 1', the well-established options, like on-land wind and large PV solar, with costs falling and support needs reducing. So only £50 million p.a. was allocated for this. Secondly, 'Pot 2', with £155 million p.a., for less-established options like offshore wind, geothermal, AD biomass, wave and tidal stream projects, which still needed more support to get prices down, some more than others. Offshore wind was amongst the cheapest of the Pot 2 group, with the recommended CfD strike price set at £155/MWh for 2014–2017, falling to £140 thereafter, and at that time costs seemed likely to fall to around £100/MWh after 2020. By contrast, the 'reference' strike price for wave and tidal stream (provided as an indicative administrative guide to what was might be acceptable) was set at £305/MWh over the whole period up to 2019.

As noted above, a 100 MW tranche was set aside for wave and tidal stream in Round 1, so arguably, if that was taken up, there would be less room (and cash) for the others. More significantly, Gordon Edge, then director of policy at Renewable UK, noted that, even if offshore wind got the full £155 million p.a. allocated to 'Pot 2', this would fund just one typical 500 MW offshore wind farm 'which is significantly less than we need'. Given, for example, the then recent planning go-ahead for E.ONs

700 MW Rampion project off Sussex, and the many others in the pipeline, that seemed quite an understatement.

It was even tighter in Pot 1. With only £50 million available per annum for hydro, energy from waste, onshore wind, landfill gas, sewage gas and large-scale solar, the STA complained that 'even if all of this went to solar – which it won't – this is only enough for 1 GW of solar in this round, a considerable reduction on the current market' (STA 2014).

The reference CfD strike price offered for PV was £120/MWh for 2014–2016, falling in stages to £100 by 2018, but PV would be in competition with onshore wind, with a reference strike price of £95/MWh falling to £90 by 2019, and with landfill and sewage gas, at even lower strike prices (£55 and 75/MWh, respectively). The STA said that, overall, the cap could cut large-scale solar installations by about 65–80%.

Imposing caps in an expanding field was bound to lead to conflicts amongst the participants, each with special cases to make. However, there were also some possible strategic contradictions from the government's point of view. At that time (i.e. in the run up to the CfD Round 1 auction), onshore wind was still allowed in the CfD and was doing well. If large solar was squeezed out, wind might do even better. But that was surely not what the Conservatives wanted. They had talked of constraining on-land wind, reflecting pressure from their (anti-wind farm) political base in the shires. Then again, the £50 million p.a. cap for Pot 1 might achieve that. Even if on-land wind got it all (so killing-off large solar), the rate of growth of on-land wind would still be cut: it was suggested that it might be halved (Toke 2014).

In the event, these various potential conflicts were avoided to some extent in CfD Round 1 by the government's decision to increase the overall CfD cap level to £300 million p.a. (DECC 2014f). So there was a bit more flexibility available, and some headroom was also allowed: the final Round 1 total allocation actually came to £315 million (DECC 2015a). As noted earlier (Box 6.5), 15 onshore wind projects got contracts, along with 5 solar projects, although, in terms of capacity, offshore wind dominated.

Fewer battles occurred in relation to Round 2 of the CfD in 2017, which was limited just to Pot 2, less-developed projects, so that, crucially, large solar and onshore wind were both now excluded from the auction.

So there was less potential for conflict over the £290 million p.a. allocation—a big increase for Pot 2 from the £155 million initially allocated in the previous round. As we have seen, offshore wind did very well, with some very low price bids winning (BEIS 2017b). That may have been good news, but it did not resolve the problem of what to do about onshore wind and PV and also what to do about the less-developed renewables.

Although the special protected tranche for wave and tidal projects that has been introduced (but not used) in Round 1 had been removed, there had been hopes that the ongoing Meyen tidal stream project in Pentland Firth and the proposed Swansea tidal lagoon might get CfD contracts at some point, but as already noted, that has not yet proved to be the case. Some already RO-contracted onshore wind projects were still starting up, as were some of the last CfD-supported onshore wind and PV projects, and with PV cost falling, domestic PV, although slowed by the cuts, was still expanding under the FiT. However, new large-scale PV and onshore wind projects were left with the option of 'going it alone' without subsidy. One 10 MW 'zero subsidy' PV project did get going (at Clayhill in Bedfordshire), and a much larger one (350 MW) was proposed for Kent (Buckley 2018). There was also talk of onshore wind doing the same, with hopes being raised that this might be a breakthrough (Vaughan 2018; Eddie 2018). Nevertheless, 'going it alone' would not be easy, especially when competing with still-subsidised projects, and the pace and scale of expansion in this mode was likely to be low, as BEIS has recognised, particularly for smaller projects (BEIS 2018c).

Clearly weaning renewables off subsides might take time, and it was argued that a 'tapered' reduction approach to subsidies would make more sense than sudden, abrupt, drastic cuts. A Policy Exchange report, 'Powering Up: The future of onshore wind in the UK', had argued that, rather than being excluded from the CfD, onshore wind should continue to receive Contracts for Difference, albeit with the subsidies phased out in stages, so that they effectively become 'subsidy free' by 2020. It predicted that this would help the onshore wind industry reduce costs from £85/MWh to approximately £60/MWh by 2020, putting it on a par with new gas plants (Howard and Drayson 2015).

The STA made a similar case for PV solar. In response to the DECC Review on the future of the FiTs, it said, 'a sudden cut combined with the threat of scheme closure is a particularly bad idea – it will create a huge boom and bust that is not only very damaging to solar businesses and jobs but does nothing to help budget constraints. We really are astonished at how self-defeating these proposals are' (STA 2015a). Its 'Solar Independence' report outlined a plan 'for a stable glide path to subsidy-free solar' (STA 2015b).

As can be seen, the funding battles had become quite bitter, and, although it was clear that cuts had to be made, the pace of change was seen as too rapid, especially for PV. By 2016, it had reached around 12 GW. Was that to be the end of it? The savings involved from some of the cuts and programme changes also seemed marginal. For example, at one point it was claimed that one of the proposed PV cuts might only save consumers as little as 50 pence per year (Macalister 2015). A proposal to end access for onshore wind projects to the RO a year early was claimed to be likely to save just 30 pence p.a. on bills. Some of the arguments used against onshore wind were also rather weak. For example, in December 2014, Energy Minister Amber Rudd said 'Onshore wind has been an important part of the mix and, of course, we have more onshore wind in this country than in the rest of the world, so I think that it may be time for us to spend our scarce resources on other types of renewables to ensure the best return for taxpayers' (Rudd 2014). In fact, at that point, globally, wind was at 360 GW, nearly all onshore. China had 91 GW, the USA 60 GW, Germany 35 GW and the UK 8 GW.

However, it was also the case that some developers had got used to having access to subsidies almost as a right. At an energy industry conference in 2015, Tim Rotheray, of the Association for Decentralised Energy, while noting that the speed of the subsidy cuts had 'made everyone nervous', said that 'there has been a sense of entitlement among some in the renewables industry', with some saying in effect 'You can't do this to me because I'm green.' This view of entitlement was reflected by Angela Knight, former CEO of Utility trade body Energy UK, who said 'if you rely on government subsidies, you are in the hands of governments. You cannot assume you have a God-given right to a subsidy' (Ross 2015).

The debate over the details of the subsidies suddenly become less relevant when, in an annex to the 2017 Budget, details were provided of a plan for replacing the Levy Control Framework, with a new 'Control of Carbon Levies' system (HM Government 2017). It would cover the RO, FiT and the Contracts for Difference systems as before, but, crucially, to keep future costs down, on the basis of current forecasts, it was noted 'there will be no new low carbon electricity levies until 2025'. The government's Budget statement annex said that the new system 'does not seek to cap or set a budget for low carbon electricity levies', but it 'will not allow for new low carbon electricity levies to be introduced'. So it amounted to a future funding freeze.

It certainly looked bleak. The existing FiT cuts for small solar, the block on onshore wind and large solar and the other constraints already in place would cause spending on renewables to level off well before 2025, as BEIS data showed, but now no new funding would be allowed, although the BEIS forecast did include a new £557 million CfD allocation. That had been proposed in the government's Budget statement, and was confirmed later (BEIS 2018e).

The new CfD allocation was something of a surprise, given the plan to freeze the LCF successor for new projects. It was likely to be mostly for offshore wind (1–2 GW p.a. was mentioned) and, allegedly, could avoid breaching the cap, since, the government said, the 'significant cost reductions that were achieved in the last CfD auction indicate that this support could secure far more low carbon electricity than originally anticipated'. The freeze was just for *new* CfDs after that, or for support under other such schemes. Nevertheless, since it had already been agreed, the Hinkley nuclear CfD would still go ahead, from whenever Hinkley started up, with, as noted before, 2027 being the current guesstimate.

With that significant £92.5/MWh nuclear commitment impending, and likely to rise by the time it was taken up (since, as noted above, the award was inflation index linked), it is perhaps clear why BEIS wanted to get the rest of the spending under control in the run up period to Hinkley's start, even if any new renewable CfD contracts involved would presumably be for much less cost than Hinkley, as was likely to be the case for the new CfD round that has been let through.

However, whatever the reason, although the door to new funding seemed to have closed, there were let-out clauses which might let some projects through: 'New low carbon electricity levies may be considered if the aggregate of existing levies is forecast to have a sustained and significant fall in real terms. Given the volatility of the forecasts and their sensitivity to changes in wholesale prices, the government would look closely at the drivers and sustainability of any decline before considering possible additional levies. Even if this condition is not satisfied, in order to ensure the lowest costs for consumers, new levies may still be considered where they have a net reduction effect on bills and are consistent with the government's energy strategy' (HM Government 2017).

So there was the possibility of some adjustment. Nevertheless, as it stood, the government indicated that support for renewables would tail off and then more or less flat-line after 2023, see Fig. 6.9, much as had been predicted, based on DECC's pessimistic projections, in 2013: see Fig. 6.2.

Even so, the government still suggested that, by 2024–2025, there would be 41.67 GW of large renewables in place, including 13.3 GW of onshore wind (its growth being static from 2019) and 14 GW of offshore wind, plus 7.2 GW of large PV solar, along with 6.1 GW of small-scale renewables, almost 5 GW of that being small PV. So nearly 48 GW in all. That was what was behind the scenario reproduced earlier in Fig. 6.5.

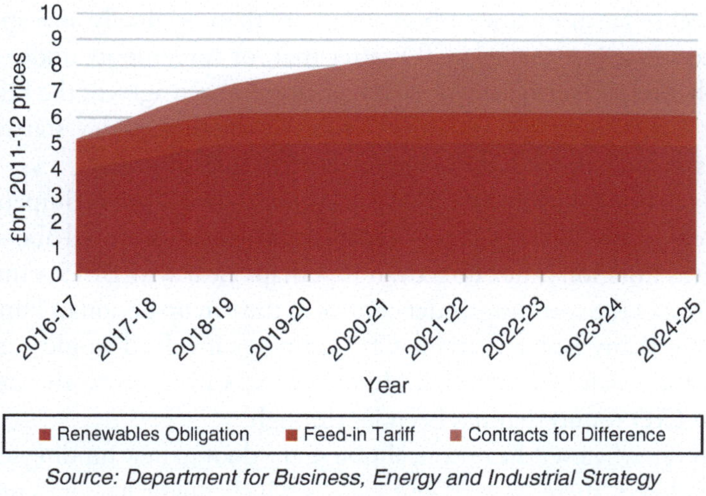

Fig. 6.9 BEIS LCF spending forecast. (HM Government 2017)

Quite apart from the question of whether all of that would be achieved, given the cuts, there was also the question of what would happen next, after 2025. The projection in Fig. 6.5 did run up to 2035, with some significant growth still occurring. By 2018, around 40 GW had been installed, so continued growth on the scale that scenario envisaged might be possible. Indeed, the BEIS spreadsheet reference scenario has 68 GW of renewables in place by 2035, with 1–2 GW being added in most years (BEIS 2018b, Annex L). However, on current policy trends, that is uncertain and would arguably require more support. The case for providing that had perhaps inadvertently been strengthened by the high CfD strike price given to Hinkley. At one time the Hinkley CfD level was, comparatively, lower than that for offshore wind and solar, but now with prices for them much lower, the Hinkley allocation looked dangerously like providing a justification for more support for renewables.

We will be coming back to look some more at the political debates over the cuts in Chap. 8, but it was clear that not all was well with the programme, as was indicated by the news that investment in UK renewables had fallen by 56% between 2006 and 2017, much more that the 26% fall that had been recorded for the EU as whole (BNEF 2018). This was despite the expansion of UK offshore wind. Evidently, for the rest, the cuts and uncertainties had undermined investor confidence. The chair of the Environment Audit Committee (EAC) said the 'dramatic fall in investment is threatening the government's ability to meet legally binding climate change targets' (Hirtenstein 2018). Something evidently needed to be done, with the EAC calling for action 'to secure the investment required to meet the UK's climate change targets' (EAC 2018).

6.5 The Next Phase

In its 2017 Clean Growth Strategy, part of its new wider Industrial Strategy (BEIS 2017e), the Conservative Governments had set the overriding policy that 'every action that the government takes to cut emissions must be done while ensuring our economy remains competitive', so the emphasis was to be on low-cost options, which made its continued commitment to nuclear a little odd, but otherwise it seemed generally sensible, although not wildly ambitious. It was admitted that the UK was on track to miss the fourth and fifth carbon budgets. However, the headline commitments looked appealing. In all, it said over £2.5 billion of investment would be allocated to support low-carbon innovation from 2015 to 2021, although, in fact, some of this had already been allocated or announced (BEIS 2017f). The overall allocations, including the cost of the new capacity market (which we will look at in the next chapter), are shown in Table 6.6.

As noted above, a new CfD round would go ahead (not shown in the Table). The aim was to provide 'up to half a billion pounds for further Contract for Difference auctions for less established technologies, such as offshore wind, with the next one planned for spring 2019', and 'work with industry as they develop an ambitious Sector Deal for offshore wind, which could result in 10 gigawatts of new capacity, with the opportunity for additional deployment if this is cost effective, built in the 2020s'. It said 'This would mean up to £557 million for further Pot 2 Contract for Difference auctions, with the next one planned for spring 2019' (BEIS 2017f).

The Environmental Audit Committee was not convinced this plan was enough, but, with at least a degree of certainty having been established about the short term, and despite continuing worries about falling inward investment, the debate then moved on to the longer term, and inevitably turned to some extent on what might happen with nuclear power. In the medium term, it did look odd to some to ignore onshore wind while funding Hinkley with a lavish CfD.

The Public Accounts Select Committee had expressed concern about the impact of the latter on consumers' power costs, at a time when renewable costs were falling dramatically (PAC 2017), while the Energy and Climate Intelligence Unit claimed that 1 GW of onshore wind could cost £1 billion less than an equivalent nuclear programme, and using onshore wind would be £100 million a year cheaper than doing so from new nuclear reactors: 'The effective ban on the cheapest form of new power generation looks increasingly perverse' (ECIU 2017).

It could also be argued that, if onshore wind and PV were put back fully on board, then, along with more offshore wind, the UK could perhaps be on the way to getting at least 50% of its power from renewables

Table 6.6 Government forecast of expenditure 2016–2021

£ billion (nominal prices)	2015–2016	2016–2017	2017–2018	2018–2019	2019–2020	2020–2021
Feed-in Tariffs	1.3	1.5	1.7	1.9	2.1	2.2
Contracts for difference	0.0	0.2	0.5	1.1	2.2	2.8
Renewable heat incentive	0.42	0.64	0.78	0.90	1.01	1.15
Capacity market	0.0	0.0	0.0	0.6	1.0	1.1

OBR (2015)

in the next decade or so, and possibly much more later. Support for that contention came from Scotland.

As noted earlier, the Scottish devolved government had adopted very radical policy, aiming to meet 100% of its electricity consumption from renewables by 2020. Its interim target was 24% by 2010, which it easily surpassed, reaching around 35% by 2011 (Scottish Government 2012). By 2017, it had reached over 60%, compared to around 25% in the UK as a whole. Clearly, very high renewable percentages were possible in some locations. Scotland was particularly blessed, with the EU's best wind resource and also an SNP-led government very keen to develop it rapidly and, if necessary, independently, even if that would require some significant changes (Wood and Baker 2017). Interestingly, Scotland was also doing well on smaller-scale local energy projects, with, by 2017, 666 MW installed, 81 MW of it being local community owned, aided by its Community and Renewables Energy Scheme. By contrast, in the rest of the UK, only 168 MW had been installed (Community Energy England 2018). The Scottish local renewables target was 1 GW by 2020.

While some were not convinced that Scotland could meet its very ambitious overall 100% renewables target, it did have a major wind resource, as well as hydro, and was also well placed for wave and tidal stream projects, which it had been developing, in part using its own funding schemes. While Scotland might therefore be seen as, in some ways, an exceptional case, scenarios emerged suggesting that the UK as a whole could do something similar. For example, Germany's *Energy Watch Group* and Finland's LUT (Lappeenranta University of Technology) produced a '100% renewable power by 2050' scenarios for 9 global regions and some countries within them, including a 100% renewables scenario for the UK/Ireland. In the latter, wind energy, on- and offshore, would increasingly drive most of the system. Its share would increase to about 59% of total power in 2050, as it became the least-cost energy source. The EWG/LUT modelling indicated that the power system LCOE would 'decline from 68.6 €/MWh to 47.7 €/MWh from 2015 to 2050, including all generation, storage, *curtailment* and parts of the grid costs. Beyond 2030 the LCOE further declines to 47.7 €/MWh by 2050, signifying that larger capacities of RE addition result in a reduction of energy costs. After an initial increase, the investment requirements decline after 2040' (EWG/LUT 2017).

Ambitious scenarios like that open up many issues, including that of grid balancing, which we will looking at in more detail later. The EWR/LUT study relied on storage, the cost of which they say contributed substantially (34%) to the overall system cost, with batteries seen as the main element for backing up PV on a daily basis (around 15 TWh p.a. output by 2050), while significant gas storage would be installed, mainly to provide 'seasonal' (i.e. longer-term) storage for winter and long wind lulls (nearly 16 TWh output capacity by 2050), using Synthetic Natural Gas (6%) and bio-methane (94%).

There have been other interesting 'high renewables' UK scenarios produced by academics, with detailed modelling of balancing needs (Pfenninger and Keirstead 2015), along with my own earlier efforts, using the open access DECC modelling system (Pugwash 2013). In 2015, an EPSRC-funded *Realising Transition Pathways Research Consortium* of nine UK universities developed transition pathways to 2050, including one with a high level of renewables and local generation (RTP 2015). Still very active in the field, the CAT in Wales has also produced a series of 'Zero Carbon Britain' high renewables scenarios (CAT 2018). Although not without its limits and problems, DECC's 2050 scenario modelling system, developed during David Mackay's tenure as DECC's Chief Scientist, certainly helped many groups, including NGOs like CAT, to produce some interesting scenarios, and it is it still available (BEIS 2018f).

However, from the mainstream, it is interesting that, in 2017, the UK's National Grid company, which is responsible for managing the UK power system, also produced a range of scenarios, including the quite ambitious one for the UK shown in Fig. 6.10.

This assumes a decentralised approach, with 'prosumer' self-generation by domestic consumer spreading, as has happened in Germany. There was 63 GW of solar and 65 GW of wind on the system by 2050 in National Grid's 'consumer renewables' scenario, with nuclear being cut back to 5 GW. In its other more conventional scenarios, nuclear grows more (to 20 GW in its '2 degrees' scenario) and renewables less, but, overall, National Grid looks to renewables having 60% of total capacity

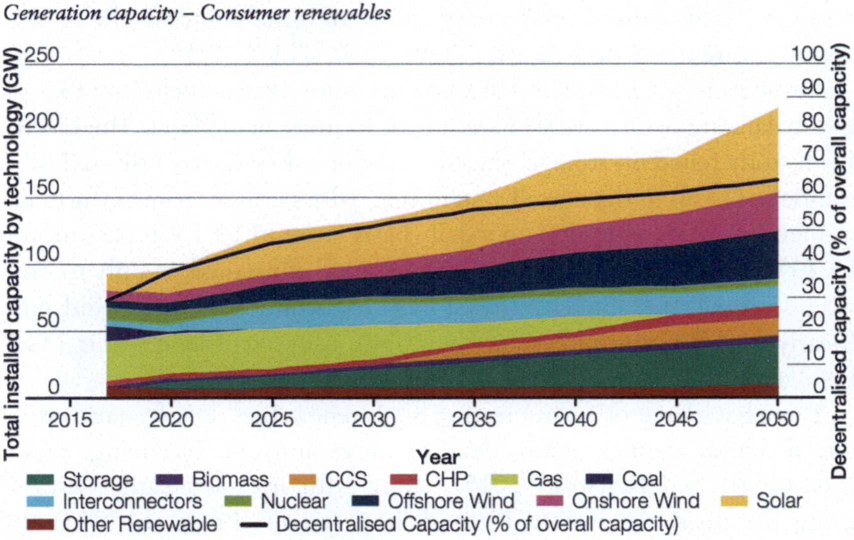

Fig. 6.10 National Grid consumer-led scenario, 2017 version. (FES 2017, © National Grid plc, all rights reserved)

by 2050, and distributed energy (including local CHP) reaching 93 GW by then. In a 2018 update of the community renewables scenario, given their cost reductions, renewables accelerate to supply 75% of power, with wind expanding to 83 GW by 2050, while the role of green hydrogen, as a new energy transmission and storage vector, is increased, some of this being for heating, some for vehicle use, issues we will be looking at in Chap. 7 (FES 2018).

Leaving speculative ideas like that aside for now, our concern here is what pathways are seen as likely by the various UK agencies involved with shaping UK policy. National Grid is something of an exception in producing ambitious scenarios. ETSU, which it will be recalled played a major assessment and advisory role in the early days, was privatised in 1996 (it became part of AEA Technology, spun off from the UK Atomic Energy Authority) and in effect mostly left the scene, with the nearest to a replacement being the hybrid private-public funded ETI, which was set up in 2007. The UK Energy Research Centre (UKERC), which was set up in 2004, at the suggestion of the then chief scientist, Sir

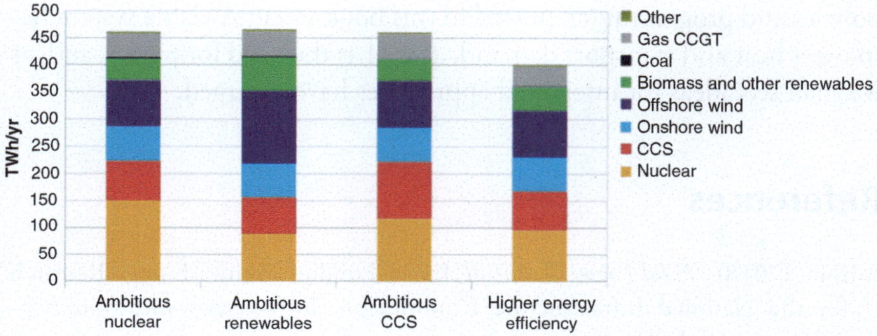

Fig. 6.11 CCC scenarios for reaching 50 gCO_2/kWh by 2030. (CCC 2013)

David King, with nodes in a number of Universities, has also continued to produce incisive technology strategy and assessment reports, as has the Carbon Trust, which was set up in 2001. For example, the Trust worked on an influential Renewables Innovation Review with the DTI in 2004. That suggested that if renewables were all developed successfully, they might supply between 53% and 67% of UK electricity by 2050 (DTI 2004).

Modelling work has also been done for the government's advisory Committee on Climate Change (CCC) and, based on that, CCC has looked to significant expansion of renewables. However, it has also backed nuclear and fossil CCS. See Fig. 6.11 for some of its scenarios.

Events since they were produced, with CCS stalled and nuclear expansion uncertain, led CCC, in its 2018 progress report, to note that 'if new nuclear projects were not to come forward, it is likely that renewables would be able to be deployed on shorter timescales and at lower cost' (CCC 2018). That was a significant shift in view, illustrating how rapidly the relative prospects for renewables and nuclear were changing.

However, not everything may change rapidly. Significant subsidy cuts have been made for renewables, but not for nuclear. It seemed to be protected. There was still no sign of a level playing field. Moreover, there were also other constraints which might make it hard for renewables to expand rapidly, including the need to balance their variability. That has surfaced as a key issue, with some of the solutions involving a move beyond a focus just on electricity, which has been the main concern of the policies and programmes reported in this book so far. A viable system has to meet heat and transport demand, as well as the need for power, and, as we shall see, ideas for integrated approaches have emerged.

References

Aurora. (2018). *Tidal Power Technical Annex*. London: Aurora Energy Research for the National Infrastructure Commission. https://www.nic.org.uk/wp-content/uploads/Tidal-power.pdf

Barker, G. (2014, April 4). Quoted in Carrington, D. (2014). Solar Farms "Will Not Spread Unrestricted Across British Countryside". *The Guardian*. https://www.theguardian.com/environment/2014/apr/04/solar-farms-spread-unrestricted-british-countryside-greg-barker-wind

BEIS. (2017a). '*Energy and Climate Change Public Attitude Tracker*', *Wave 23*. London: Department for Business, Energy and Industrial Strategy. https://assets.publishing.service.gov.uk/government/uploads/system/uploads/attachment_data/file/656549/Wave_23_Summary_Report.pdf

BEIS. (2017b). *Contracts for Difference Second Allocation Round Results*. London: Department for Business, Energy and Industrial Strategy. http://www.gov.uk/government/uploads/system/uploads/attachment_data/file/643560/CFD_allocation_round_2_outcome_FINAL.pdf

BEIS. (2017c). *Solar Photovoltaics Deployment*. London: Department for Business, Energy and Industrial Strategy. http://www.gov.uk/government/statistics/solar-photovoltaics-deployment

BEIS. (2017d). *Review of Support for Anaerobic Digestion and Micro-Combined Heat and Power Under the Feed-in Tariff Scheme: Government Response*. London: Department for Business, Energy and Industrial Strategy. http://www.gov.uk/government/consultations/review-of-support-for-anaerobic-digestion-and-micro-combined-heat-and-power-under-the-feed-in-tariffs-scheme

BEIS. (2017e). *Industrial Strategy: Building a Britain Fit for the Future*. London: Department for Business, Energy and Industrial Strategy. https://www.gov.uk/government/news/government-unveils-industrial-strategy-to-boost-productivity-and-earning-power-of-people-across-the-uk

BEIS. (2017f). *Clean Growth Strategy*. London: Department for Business, Energy and Industrial Strategy. https://www.gov.uk/government/publications/clean-growth-strategy

BEIS. (2018a). *Energy and Climate Change Public Attitude Tracker, Wave 25*. London: Department for Business, Energy and Industrial Strategy. https://assets.publishing.service.gov.uk/government/uploads/system/uploads/attachment_data/file/702640/Wave_25_Summary_Report.pdf

BEIS. (2018b). *Updated Energy and Emissions Projections 2017*. London: Department for Business, Energy and Industrial Strategy. https://assets.publishing.service.gov.uk/government/uploads/system/uploads/attachment_data/file/671187/Updated_energy_and_emissions_projections_2017.pdf

BEIS. (2018c). *The Future for Small-Scale Low-Carbon Generation*. London: Department for Business, Energy and Industrial Strategy. https://assets.publishing.service.gov.uk/government/uploads/system/uploads/attachment_data/file/727138/Call_for_evidence-Future_SSLCG.pdf

BEIS. (2018d, June 25). *Proposed Swansea Bay Tidal Lagoon*. Business, Energy and Industrial Strategy Secretary Greg Clark's Statement to Parliament. https://www.gov.uk/government/speeches/proposed-swansea-bay-tidal-lagoon

BEIS. (2018e). *Energy Minister Claire Perry Hails Success Story of Offshore Wind in Newcastle Today*. London: Department for Business, Energy and Industrial Strategy. https://www.gov.uk/government/news/energy-minister-claire-perry-hails-success-story-of-offshore-wind-in-newcastle-today

BEIS. (2018f). *DECC 2050 Scenario Tool*. London: Department for Business, Energy and Industrial Strategy. https://www.gov.uk/guidance/2050-pathways-analysis

BERR. (2008). *Energy White Paper: Meeting the Energy Challenge*. London: Department of Business, Enterprise & Regulatory Reform. https://assets.publishing.service.gov.uk/government/uploads/system/uploads/attachment_data/file/228944/7296.pdf

Blair, T. (2006, May 16). CBI Speech, as Quoted by the BBC, *Blair Backs Nuclear Power Plans*. http://news.bbc.co.uk/1/hi/uk_politics/4987196.stm

BNEF. (2018, January 16). Bloomberg New Energy Finance Press Release. https://about.bnef.com/blog/runaway-53gw-solar-boom-in-china-pushed-global-clean-energy-investment-ahead-in-2017/

Brack, D. (2017). *Woody Biomass for Power and Heat: Impacts on the Global Climate*. London: Chatham House. http://www.chathamhouse.org/publication/woody-biomass-power-and-heat-impacts-global-climate

Buckley, A. (2018, July 19). How a Solar Farm in Southeast England Could Bring a New Dawn for Renewable Energy. *The Conversation*. https://theconversation.com/how-a-solar-farm-in-southeast-england-could-bring-a-new-dawn-for-renewable-energy-99530

Business Green. (2017, November 24). German Auction Sees Wind Energy Costs Drop Another 10 Per cent. *Business Green*. http://www.businessgreen.com/bg/news/3021838/german-auction-sees-wind-energy-costs-drop-another-10-percent

Butler, G., Evans, M., Inderwildi, O., King, D., & McGlynn, G. (2012, March). *Towards a Low Carbon Pathway for the UK*. Smith School of Enterprise and the Environment, University of Oxford. https://www.smithschool.ox.ac.uk/publications/reports/towards-a-low-carbon-pathway-for-the-uk-report-2012.pdf

CAB. (2017). *Generating Value?* London: Citizens Advice Bureau. https://www.citizensadvice.org.uk/Global/CitizensAdvice/Energy/GeneratingValue.pdf

Cameron, D. (2016, January 12). *House of Commons Liaison Committee, Oral Evidence: Evidence from the Prime Minister, Q 83, HC 712*. http://data.parliament.uk/WrittenEvidence/CommitteeEvidence.svc/EvidenceDocument/liaison/evidence-from-the-prime-minister-12-january-2016/oral/26761.htm

CAT. (2018). *A Series of Zero Carbon Britain Reports*. Centre for Alternative Technology Reports. http://www.zerocarbonbritain.org/en/

CCA. (2008). Climate Change Act. http://www.legislation.gov.uk/ukpga/2008/27/contents

CCC. (2013). *Next Steps on Electricity Market Reform – Securing the Benefits of Low-Carbon Investment*. London: Committee on Climate Change. https://www.theccc.org.uk/wp-content/uploads/2013/05/1720_EMR_report_web.pdf

CCC. (2018). *Reducing UK Emissions – 2018 Progress Report to Parliament*. London: Committee on Climate Change. http://www.theccc.org.uk/publication/reducing-uk-emissions-2018-progress-report-to-parliament/

Community Energy England. (2018). *Community Energy: State of the Sector 2018*. Community Energy England/Community Energy Wales. https://communityenergyengland.org/files/document/169/1530262460_CEE_StateoftheSectorReportv.1.51.pdf

DECC. (2010a). *Feed in Tariff Impact Assessment (and Consultation Results)*. London: Department of Energy and Climate Change. http://webarchive.nationalarchives.gov.uk/20121217165258/http://www.decc.gov.uk/assets/decc/consultations/renewable%20electricity%20financial%20incentives/2710-final-ia-feed-in-tariffs-small-scale.pdf

DECC. (2010b). *Severn Tidal Power: Feasibility Study Conclusions and Summary Report*. London: Department of Energy and Climate Change. http://www.decc.gov.uk/assets/decc/What%20we%20do/UK%20energy%20supply/

Energy%20mix/Renewable%20energy/severn-tp/621-severn-tidal-power-feasibility-study-conclusions-a.pdf
DECC. (2011a). *UK Renewable Energy Roadmap*. London: Department of Energy and Climate Change. https://assets.publishing.service.gov.uk/government/uploads/system/uploads/attachment_data/file/48128/2167-uk-renewable-energy-roadmap.pdf
DECC. (2011b). *Planning Our Electric Future: A White Paper for Secure, Affordable and Low-Carbon Energy*. White Paper, Department of Energy and Climate Change, London. https://www.gov.uk/government/publications/planning-our-electric-future-a-white-paper-for-secure-affordable-and-low-carbon-energy
DECC. (2011c). *Barker: Boom and Bust for Solar Must be Avoided*. London: Department of Energy and Climate Change. https://www.gov.uk/government/news/barker-boom-and-bust-for-solar-must-be-avoided
DECC. (2012a). *DECC Public Attitudes Tracker, Wave 1*. London: Department of Energy and Climate Change. https://assets.publishing.service.gov.uk/government/uploads/system/uploads/attachment_data/file/48446/5707-decc-public-att-track-surv-wave1-summary.pdf
DECC. (2012b). *The Energy Bill: Electricity Market Reforms*. London: Department of Energy and Climate Change. http://www.decc.gov.uk/en/content/cms/news/pn12_151/pn12_151.aspx
DECC. (2012c). *UK Renewable Energy Roadmap Update*. London: Department of Energy and Climate Change. https://assets.publishing.service.gov.uk/government/uploads/system/uploads/attachment_data/file/80246/11-02-13_UK_Renewable_Energy_Roadmap_Update_FINAL_DRAFT.pdf
DECC. (2013a). *UK Solar PV Strategy Part 1: Roadmap to a Brighter Future*. London: Department of Energy and Climate Change. https://www.gov.uk/government/publications/uk-solar-pv-strategy-part-1-roadmap-to-a-brighter-future
DECC. (2013b). *Energy and Emissions Update*. London: Department of Energy and Climate Change. https://assets.publishing.service.gov.uk/government/uploads/system/uploads/attachment_data/file/239937/uep_2013.pdf
DECC. (2014a). *UK Solar PV Strategy Part 2*. London: Department of Energy and Climate Change. https://assets.publishing.service.gov.uk/government/uploads/system/uploads/attachment_data/file/302049/uk_solar_pv_strategy_part_2.pdf
DECC. (2014b). *Electricity Market Reform: Allocation of Contracts for Difference. A Government response on Competitive Allocation*. London: Department of

Energy and Climate Change. https://assets.publishing.service.gov.uk/government/uploads/system/uploads/attachment_data/file/310272/competitive_allocation_government_response.pdf

DECC. (2014c). *Electricity Market Reform: Contracts for Difference*. London: Department of Energy and Climate Change. http://www.gov.uk/government/publications/electricity-market-reform-contracts-for-difference

DECC. (2014d). *Electricity Market Reform: Allocation of Contracts for Difference* (p. 13). London: Department of Energy and Climate Change. https://www.gov.uk/government/uploads/system/uploads/attachment_data/file/271919/Competitive_allocation_consultation_formatted.pdf

DECC. (2014e). *Indicative CfD Budget Notice for the Autumn 2014 CfD Allocation Round*. London: Department of Energy and Climate Change. https://www.gov.uk/government/news/over-200-million-boost-for-renewables

DECC. (2014f). *£300 Million Budget to Launch UK Auctions for Renewables*. London: Department of Energy and Climate Change. https://www.gov.uk/government/news/300-million-budget-to-launch-uk-auctions-for-renewables

DECC. (2015a). *Contracts for Difference (CFD) Allocation Round One Outcome*. London: Department of Energy and Climate Change. https://www.gov.uk/government/publications/contracts-for-difference-cfd-allocation-round-one-outcome

DECC. (2015b). *Review of the Feed-in Tariffs Scheme*. London: Department of Energy and Climate Change. http://www.gov.uk/government/uploads/system/uploads/attachment_data/file/486082/FITs_Review_Govt__response_Final.pdf

DECC. (2015c). *Consultation on a Review of the Feed-in Tariffs Scheme*. London: Department of Energy and Climate Change. https://assets.publishing.service.gov.uk/government/uploads/system/uploads/attachment_data/file/469476/Consultation_on_a_Review_of_feed-in_tariff_scheme.pdf

DECC. (2015d). *EMR and FID Enabling for Renewables Government Response*. London: Department of Energy and Climate Change. https://www.gov.uk/government/publications/electricity-market-reform-evaluation

DEFRA. (2014). *Subsidies for Solar Farms to be Cut to Help Safeguard Farmland*. London: Department for Environment, Food and Rural Affairs. https://www.gov.uk/government/news/subsidies-for-solar-farms-to-be-cut-to-help-safeguardfarmland

DTI. (2003). *Energy White Paper 2003: Our Energy Future – Creating a Low Carbon Economy*. London: Department of Trade and Industry. http://webarchive.nationalarchives.gov.uk/+/www.berr.gov.uk/files/file10719.pdf

DTI. (2004). *Renewables Innovation Review*. London: Department of Trade and Industry/Carbon Trust. http://webarchive.nationalarchives.gov.uk/20060215070818/http://www.dti.gov.uk/renewables/renew_2.1.4.htm
DTI. (2005). *Renewables Explained*. London: Department of Trade and Industry. http://webarchive.nationalarchives.gov.uk/20060214073333/http://www.dti.gov.uk/renewables/renew_1.1.2.2.htm
DTI. (2006a). *The Energy Challenge*. London: Department of Trade & Industry. https://assets.publishing.service.gov.uk/government/uploads/system/uploads/attachment_data/file/272376/6887.pdf
DTI. (2006b). *Our Energy Challenge: Power from the People, Microgeneration Strategy*. London: Department of Trade & Industry. http://webarchive.nationalarchives.gov.uk/20090203191946/http://www.berr.gov.uk/whatwedo/energy/sources/sustainable/microgeneration/strategy/page27594.html
DTI. (2007). *Meeting the Energy Challenge: A White Paper on Energy*. London: Department of Trade & Industry. https://assets.publishing.service.gov.uk/government/uploads/system/uploads/attachment_data/file/243268/7124.pdf
EAC. (2006). *Keeping the Lights On: Nuclear, Renewables and Climate Change*. Environmental Audit Committee, Sixth Report of Session 2005–06. https://publications.parliament.uk/pa/cm200506/cmselect/cmenvaud/584/584i.pdf
EAC. (2018). *Green Financing Report*. Environmental Audit Select Committee. http://www.parliament.uk/business/committees/committees-a-z/commons-select/environmental-audit-committee/news-parliament-2017/green-finance-report-published-17-19/
Early, C. (2018, August 7). Can Marine Renewables Ever Be Mainstream? *Utility Week*. https://utilityweek.co.uk/can-marine-renewables-ever-mainstream/
EC. (2009). *Renewable Energy Directive, and Updates*. Brussels: European Commission. https://ec.europa.eu/energy/en/topics/renewable-energy/renewable-energy-directive
ECIU. (2017). *Blown Away*. London: Energy and Climate Intelligence Unit. http://eciu.net/press-releases/2017/britain-in-1bn-block-on-cheapest-energy-technology
Eddie. (2018, March 22). *Britain on Course for £20bn Subsidy-Free Renewables 'Revolution'*. Eddie.net. http://www.edie.net/news/10/Britain-on-course-for%2D%2D20bn-subsidy-free-renewables%2D%2Drevolution-/
Elliott, D. (1994). Public Reactions to Windfarms: The Dynamics of Opinion Formation. *Energy and Environment, 5*(4), 343–362.
Elliott, D. (2003). *Energy, Society and Environment*. London: Routledge.

Elliott, D. (2007). Sea Power – How We Can Tap Wave and Tidal Power. In D. Elliott (Ed.), *Sustainable Energy*. Basingstoke: Palgrave. Online at https://link.springer.com/book/10.1057%2F9780230378384

Elliott, D. (Ed.). (2010). *Nuclear or Not?* Basingstoke: Palgrave (First Edition 2007).

Elliott, D. (2015). *Green Energy Futures*. Basingstoke: Palgrave.

Elliott, D. (2017). *Nuclear Power: Past, Present and Future*. Bristol: Institute of Physics Publications.

Elliott, D. (2018, forthcoming). Carbon Capture and Renewables. In G. Wood, et al. (Eds.), *The Long Goodbye to Fossil Fuels*. Basingstoke: Palgrave.

Elliott, D., & Wood, G. (Eds.). (2018). *Energy, Climate and Environment Book Series*. Basingstoke: Palgrave.

Energy and Climate Committee. (2012). *The Future of Marine Renewables in the UK*. House of Commons Energy and Climate Change Select Committee, Eleventh Report, Session 2011–2012, London. http://www.publications.parliament.uk/pa/cm201012/cmselect/cmenergy/1624/162402.htm

Ernst and Young. (2008). *Renewable Energy Country Attractiveness Indices*. Ernst and Young Report, Quarter 1–2.

ETI. (2017, January 16). *ETI Sets Out Priorities for Marine Energy If It Is to Compete with Other Low Carbon Sources*. Loughborough: Energy Technologies Institute. http://www.eti.co.uk/news/eti-sets-out-priorities-for-marine-energy-if-it-is-to-compete-with-other-low-carbon-sources

Europa. (2014). *Commission Concludes Modified UK Measures for Hinkley Point Nuclear Power Plant are Compatible with EU Rules*. European Commission Press Release. http://europa.eu/rapid/press-release_IP-14-1093_en.htm

Evans, S. (2018, January 8). Analysis: UK Government Slashes Outlook for New Gas Power Plants. *Carbon Brief*. https://www.carbonbrief.org/analysis-uk-government-slashes-outlook-for-new-gas-power-plants

EWG/LUT. (2017). *Global 100% RE System: Europe – UK, Ireland*. Berlin/Lappeenranta: Energy Watch Group/LUT. https://www.researchgate.net/publication/320776623_Global_100_RE_System_Europe_-_UK_Ireland

FES. (2017). *Future Energy Scenarios*, National Grid. http://fes.nationalgrid.com/media/1253/final-fes-2017-updated-interactive-pdf-44-amended.pdf

FES. (2018). *Future Energy Scenarios*, National Grid, 2018 Update. http://fes.nationalgrid.com/media/1363/fes-interactive-version-final.pdf

Grotz, C., & Fouquet, D. (2005, April). Fixed Prices Work Better. *New Energy, 2*.

Hannon, M., van Diemen, R., & Skea, J. (2017). *Examining the Effectiveness of Support for UK Wave Energy Innovation Since 2000: Lost at Sea or a New Wave of Innovation?* Imperial College London/Strathclyde University Report. https://strathprints.strath.ac.uk/62210/

Harvey, F. (2012, August 1). UK Cuts Feed-in Tariff for Solar Panels. *The Guardian*. https://www.theguardian.com/environment/2012/aug/01/solar-panel-feed-in-tariff-cut

Hendry, C. (2016). *The Role of Tidal Lagoons*. Independent Report. https://hendryreview.files.wordpress.com/2016/08/hendry-review-final-report-english-version.pdf

Hirtenstein, A. (2018, May 16). Green Investing in Britain Falls to Lowest in Decade. *Renewable Energy World*. http://www.renewableenergyworld.com/articles/2018/05/green-investing-in-britain-falls-to-lowest-in-decade.html

HM Government. (2013). *Initial Agreement Reached on New Nuclear Power Station at Hinkley*. UK Government Press Release. https://www.gov.uk/government/news/initial-agreement-reached-on-new-nuclear-power-station-at-hinkley

HM Government. (2017). *Control for Low Carbon Levies*. London: UK Government/Treasury Policy Statement. http://www.gov.uk/government/uploads/system/uploads/attachment_data/file/660986/Control_for_Low_Carbon_Levies_web.pdf

Howard, R., & Drayson, K. (2015). *Powering Up: The Future of Onshore Wind in the UK*. London: Policy Exchange. https://policyexchange.org.uk/publication/powering-up-the-future-of-onshore-wind-in-the-uk/

Macalister, T. (2015, July 22). Solar Power Subsidies Cut Might Save Just 50p on Average Electricity Bill. *The Guardian*. https://www.theguardian.com/environment/2015/jul/22/solar-power-subsidies-to-be-cut-under-plans-to-reduce-green-energy-costs

Merrill, J. (2014, October 19). Eric Pickles Accused of 'Rejecting Wind Farms to Win Votes'. *The Independent*. http://www.independent.co.uk/environment/green-living/eric-pickles-accused-of-rejecting-wind-farms-to-win-votes-9804278.html

Mitchell, C., Bauknecht, D., & Connor, P. (2006). Effectiveness Through Risk Reduction: A Comparison of the Renewable Obligation in England and Wales and the Feed-in System in Germany. *Energy Policy, 34*(3), 297–305. http://www.sciencedirect.com/science/article/pii/S0301-4215(04)00241-1

Montag, H., Parker, G., & Clarkson, T. (2016). *The Effects of Solar Farms on Local Biodiversity*. Clarkson & Woods and Wychwood Biodiversity. http://www.clarksonwoods.co.uk/download/Solar_Farms_Biodiversity_Study.pdf

Mott MacDonald. (2011, May). *Costs of Low-Carbon Technologies*. Report for the Committee on Climate Change. https://www.theccc.org.uk/archive/aws/Renewables%20Review/MML%20final%20report%20for%20CCC%209%20may%202011.pdf

Murray, J. (2014, November 5). Defra Admits It Cannot Say How Much Farmland Solar Power Is Affecting. *The Guardian*. https://www.theguardian.com/environment/2014/nov/05/defra-admits-it-cannot-say-how-much-farmland-solar-power-is-affecting

NAO. (2005). *Department of Trade and Industry: Renewable Energy*. London: National Audit Office. https://www.nao.org.uk/report/department-of-trade-and-industry-renewable-energy/

NAO. (2016). *Nuclear Power in the UK*. London: National Audit Office. http://www.nao.org.uk/report/nuclear-power-in-the-uk

NATTA. (2007). *Wave Power and Tidal Current Turbines: A Review of Progress*. Compilation of Reports from Renew, Network for Alternative Technology and Technology Assessment, Milton Keynes.

Newbery, D. (2011). *Reforming Competitive Electricity Markets to Meet Environmental Targets*. Electricity Policy Research Group, Cambridge University. https://www.eprg.group.cam.ac.uk/wp-content/uploads/2014/01/EPRG-1126_complete1.pdf

NPS. (2011). *National Policy Statements on Energy EN-1* (p. 22). London: Department of Energy and Climate Change. https://assets.publishing.service.gov.uk/government/uploads/system/uploads/attachment_data/file/47854/1938-overarching-nps-for-energy-en1.pdf

NRDC. (2016). *Money to Burn? The U.K. Needs to Dump Biomass and Replace Is Coal Plants with Truly Clean Energy*. US Natural Resources Defence Council. https://www.nrdc.org/sites/default/files/uk-biomass-replace-coal-clean-energy-ib.pdf

OBR. (2015). *Economic and Fiscal Outlook – Nov.* London: Office of Budgetary Responsibility http://obr.uk/efo/economic-and-fiscal-outlook-november-2015/

Ofgem. (2005). *Ofgem's Response to the Preliminary Consultation on the 2005–6 Review of the Renewables Obligation*. London: Office of Gas and Electricity Markets. https://www.ofgem.gov.uk/sites/default/files/docs/2007/02/16892-prelimcondoc230605.pdf

Ofgem. (2018). *Renewable Obligation Accredited Capacity*. London: Office of Gas and Electricity Markets. https://www.ofgem.gov.uk/environmental-programmes/ro/contacts-publications-and-data/public-reports-and-data-ro#thumbchart-c8935703234747052-n92708

PAC. (2014). *Early Contracts for Renewable Electricity*. London: Public Accounts Committee, House of Commons. https://www.parliament.uk/business/committees/committees-a-z/commons-select/public-accounts-committee/news/report-early-contracts-for-renewable-electricity/

PAC. (2017). *Hinkley Point C*. Public Accounts Committee, Third Report of Session 2017–19, House of Commons. https://publications.parliament.uk/pa/cm201719/cmselect/cmpubacc/393/393.pdf

Pfenninger, S., & Keirstead, J. (2015). Renewables, Nuclear, or Fossil Fuels? Scenarios for Great Britain's Power System Considering Costs, Emissions and Energy Security. *Applied Energy, 152*, 83–93. https://www.sciencedirect.com/science/article/pii/S0306261915005656

PIU. (2002). *The Energy Review*. London: Performance and Innovation Unit, Cabinet Office. http://www.gci.org.uk/Documents/TheEnergyReview.pdf

Pugwash. (2013). *High Renewable Scenario in 'Pathways to 2050: Three Possible UK Energy Strategies'*. London: British Pugwash Group. https://britishpugwash.org/pathways-to-2050-three-possible-uk-energy-strategies/

RAB. (2008). *Wave and Tidal Stream Review*. London: Renewables Advisory Board, DECC.

REA. (2017). *REA Response to Chatham House Report – Woody Biomass for Power and Heat*. London: Renewable Energy Association. http://www.r-e-a.net/news/rea-response-to-chatham-house-report-woody-biomass-for-power-and-heat

Renew. (2000). *Net Loss*. Renew 125 May–June, Renew On-line 26. http://eeru.open.ac.uk/natta/renewonline/rol26/3.htm

Renew. (2009). *Microgeneration FiT/Lords Debate on the FiT*. Renew 178 March–April, Renew On-line 78. http://eeru.open.ac.uk/renew_online.php

Ross, D. (2002). Scuppering the Waves: How They Tried to Repel Clean Energy. *Science and Public Policy, 29*(1), 25–35.

Ross, K. (2015, September 11). Renewable Subsidy Cuts Dominate UK Energy Debate. *Power Engineering International*. http://www.powerengineeringint.com/articles/2015/11/renewable-subsidy-cuts-dominate-uk-energy-debate.html

RTP. (2015). *Distributing Power: A Transition to a Civic Energy Future*. Report of the Realising Transition Pathways Research Consortium 'Engine Room'. http://www.realisingtransitionpathways.org.uk/realisingtransitionpathways/news/distributing_power.html

Rudd, A. (2014). Parliamentary Answer, House of Commons, Dec. 18th, Hansard, Vol 589. https://hansard.parliament.uk/Commons/2014-12-18/

debates/14121837000013/UKEnergySources(Subsidy)#contribution-14121837000024

Rudd, A. (2015). Parliamentary Statement, House of Commons, June 22nd, Hansard, Col. 617. http://www.publications.parliament.uk/pa/cm201516/cmhansrd/cm150622/debtext/150622-0001.htm#1506227000002

RUK. (2017, January 16). Renewable UK, Quoted in Vaughan, A., UK Wave Power Far Too Costly, Warns Energy Research Body. *The Guardian*. https://www.theguardian.com/environment/2017/jan/16/uk-wave-power-far-too-costly-warns-energy-research-body

Science and Technology Committee. (2001). *Wave and Tidal Energy*. House of Commons Select Committee on Science and Technology, Seventh Report, Session 2000–2001, London.

Scottish Government. (2012, March 29). *Scotland Beats 2011 Green Energy Target*. http://www.scotland.gov.uk/News/Releases/2012/03/geenenergytargets29032012

Shankleman, J., & Murray, J. (2014, May 15). Are Solar Farms Really Too Costly for DECC's Budget? *Business Green*. http://www.businessgreen.com/bg/analysis/2344769/are-solar-farms-really-too-costly-for-decc-s-budget

Solar Century. (2014, May 13). Seb Berry from Solar Century Quoted in McGrath, M. (2014). Government to Slash Subsidies for Large Scale Solar Farms. *BBC News*. https://www.bbc.co.uk/news/science-environment-27393805

STA. (2014, November–December). Solar Trade Association Press Release, Quoted in Renew 212, p. 3. Renew On-line 112. https://renewonlinearchive.files.wordpress.com/2015/02/rol112.pdf

STA. (2015a, August). *Solar Industry Reeling at Self-Defeating Feed-in Tariff Proposals*. London: Solar Trade Association, Undated Press Release. http://www.solar-trade.org.UK/solar-industry-reeling-at-self-defeating-feed-in-tariff-proposals

STA. (2015b). *The Solar Independence Plan for Britain*. London: Solar Trade Association. http://www.solar-trade.org.uk/solar-independence-plan-for-britain/

Telegraph. (2012, January 30). Text of a Letter Sent to PM David Cameron by 100 MPs. *Daily Telegraph*. https://www.telegraph.co.uk/news/earth/energy/windpower/9061554/Full-letter-from-MPs-to-David-Cameron-on-wind-power-subsidies.html

Tickell, O. (2015, July 9). Osborne's £3.9bn Stealth Attack on Renewables. *The Ecologist*. https://theecologist.org/2015/jul/09/osbornes-ps39bn-stealth-attack-renewables

Toke, D. (2007a). Renewable Financial Support Systems and Cost-Effectiveness. *Journal of Cleaner Production, 15*, 280–287.

Toke, D. (2007b, August). *In Making the UK Renewables Programme FITTER*. World Future Council Report.

Toke, D. (2014, July 25). *Government Cuts Onshore Wind Deployment by 50 Per cent and Solar Farms to Zero*. Green Energy Blog. http://realfeed-intariffs.blogspot.com/2014/07/government-cuts-onshore-wind-deployment.html

Treasury. (2013). *Investing in Britain's Future*. London: HM Treasury. https://www.gov.uk/government/publications/investing-in-britains-future

Vaughan, A. (2018, March 20). Subsidy-Free Renewable Energy Projects Set to Soar in UK, Analysts Say. *The Guardian*. https://www.theguardian.com/business/2018/mar/20/uk-subsidy-free-renewable-energy-projects-set-soar-aurora-energy-research-analysts

Walters, S., & Carlin, B. (2014, October 18). Eclipse of the Solar Farms. *Mail On line*. http://www.dailymail.co.uk/news/article-2798657/eclipse-solar-farms-environment-secretary-liz-truss-tells-farmers-no-handouts-ugly-fields-glass-grow-veg.html

Whitlock, R. (2016, June 2). UK Renewables Threatened Again: REM Talks to ADBA About the Threat to British Biogas. *Renewable Energy Magazine*. https://www.renewableenergymagazine.com/interviews/uk-renewables-threatened-again-rem-talks-to-20160602

Wicks, M. (2008). Parliamentary Answer, House of Commons, July 17th. Link to Hansard Entry Not Currently Available.

Wilson, B. (2001). Parliamentary Answer, Support for Renewables, Hansard 21st Nov., Col 300-01W. Holding Answer, Nov. 7th, Col. 306W. https://api.parliament.uk/historic-hansard/written-answers/2001/nov/21/renewable-energy#S6CV0375P0_20011121_CWA_200

Wood, G., & Baker, K. (Eds.). (2017). *A Critical Review of Scottish Renewable and Low Carbon Energy Policy*. Basingstoke: Palgrave.

7

Sorting the System *in the 2010s*

7.1 An Integrated System for Power, Heat and Transport

Our emphasis so far has been on electricity supply. Figure 7.1 provides a snapshot of UK progress with renewable electricity up to 2016 and projections up to 2020. However, electricity is not the only requirement. Some heating and transport demand can be met by using electricity, and as we shall see, that has been proposed, but there is also a need for renewable transport fuels and for green heat sources. This Chapter lays out the issues, and how they might be addressed, before moving to what has been said and done so far in the UK in relation to these sectors and to their integration into a balanced renewable energy system.

The basic targets, based on the overall UK commitment to get 15% of its energy from renewables by 2020, in line with the EU Renewables Directive, were to supply 30% of electricity, 12% of heat and 10% of transport fuel from renewable sources by 2020 (DECC 2010a). It has found moving towards that hard. While it had proved possible to get around 30% of UK electricity from renewables by 2018, it had proved much more difficult to expand on the heat and transport side. That is

D. Elliott, *Renewable Energy in the UK*, Energy, Climate and the Environment,
https://doi.org/10.1007/978-3-030-04765-8_7

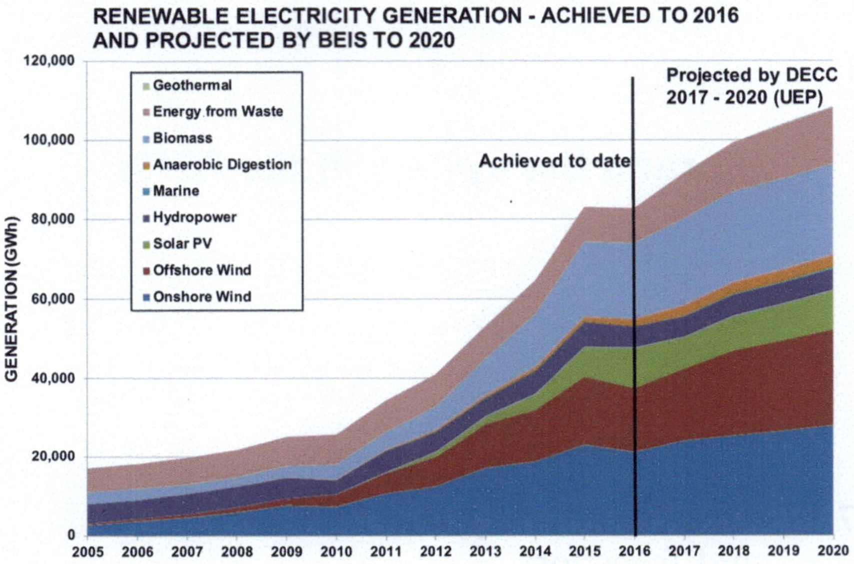

Fig. 7.1 Renewable electricity outputs based on DECC/BEIS data in 2016 and projections to 2020, under existing programmes ('UEP'). (REA 2017)

being addressed via the Renewable Heat Incentive (RHI) and Renewable Transport Fuel Obligation, with part of the focus being on the use of biomass/biogas and biofuels. See Box 7.1.

Box 7.1 The RHI and RTFO

The Renewable Heat Incentive (RHI) was set up in stages from 2011, aiming to boost the take-up rates of renewable and low-carbon heating, with much of the emphasis being on biomass heating, although solar heat and electric-fired heat pumps were also supported. There were separate grant support schemes for small-scale domestic and larger-scale commercial/industrial applications, the domestic scheme going live later, in 2014. That was targeted mainly on users who were off the gas grid and was planned to be open to new applications until March 2021. Final payments to participants for both schemes would run to at least 2040–2041, by which time the scheme was expected to have cost the taxpayer £23 billion. As we shall see later (in Sect. 7.6), there were problems with uptake. As a result, according to a critical Public Accounts Committee review published in 2018, 'the scheme is now expected to make a significantly smaller contribution to UK obligations to produce renewable heat and reduce carbon emissions, which have also been cut by 65% and 44% respectively' (PAC 2018).

The Renewable Transport Fuel Obligation initially required transport fuel suppliers to ensure that 5% of all road vehicle fuel came from sustainable renewable sources by 2010. It went into force in 2008. In practice, the Obligation has mostly been met by blending various biofuels with conventional fuel up to 5%. There have been concerns about the environmental sustainability of some of the fuels used, especially since some were imported from biofuel plantations overseas, although that proportion has been reduced. Given the land-use constraints, few however saw liquid biofuel as becoming a major vehicle fuel in the UK, although biogas produced from farm and food waste might play a role. So may might hydrogen, as we shall see later. The government has supported all these options to varying degrees via its Low Emission Vehicle programme (DoT 2018a), but the emphasis has increasingly been on supporting electric vehicle roll-out and the provision of public/private charging points.

It should also be noted that improved public transport represents part of the solution, since it is more energy efficient, and there are opportunities for using renewable sources, for example, in trains, but exploring the wider issues of transport policy (and the merits of cycling and walking!) takes us outside of the remit of this book.

Biogas is already used in some vehicles and for some heating applications, but the UK biomass resource has limits, including land-use constraints on major expansion. For example, one early study suggested that that biomass might only be able to provide about 7% of UK heating needs (DEFRA 2005). As we shall see, that might be pessimistic, but a parallel and indeed expanding focus has been on using non-fossil *electricity* for heating and transport. Indeed, some have seen that as the best way forward long term. For example, the government initially planned to meet the bulk of heating and transport demand from non-fossil electricity sources, that is, renewables and nuclear. Nuclear of course could not be counted against the '15% by 2020' EU renewable energy target, but the use of nuclear electricity could relieve renewables from meeting some heating and/or transport supply requirements. Longer term, some nuclear plants (e.g. SMRs) sited locally, might also be able to supply heat direct to users, though that is very speculative and not without risks. Certainly, for the immediate future, with the nuclear contribution relatively small

(18–20%), the use of renewable electricity (at around 30% and growing) looked likely to be the main option for the 'decarbonisation' of heat and transport.

The basic idea was to replace gas central heating with the use of electric-powered domestic-scale heat pumps (which are more efficient that direct electrical heating), and also to stimulate the use of electric vehicles (EVs), with their batteries being charged from the power grid. However, this 'decarbonisation via electrification' plan faced some problems. The power grid might not be able to deal with the significantly increased demand, especially since it was likely to peak for both uses in the evenings. The UK gas grid typically carries about four times more energy than the power grid, and services the major and very variable heat load. The daily and seasonal swings in heat demand are much larger than the variations in demand for electricity. See Fig. 7.2. Add in the large potential demand for EV battery charging (estimates varied from 10% to 20% extra), and it was clear that the power grid could face problems at times, even if sufficient generation power was actually available. There was an extended debate on the figures (Evans 2017), but most likely, the power grid, and

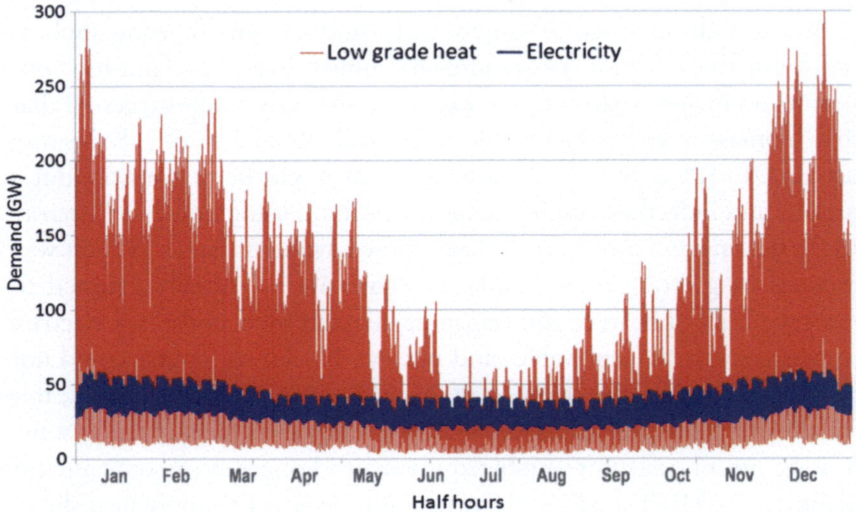

Fig. 7.2 Annual UK heat and power demand profiles in 2010. (Douglas 2015, using data from Samson and Strbac 2012)

the power generation system, would both have to be significantly upgraded and expanded.

Some of these problems might be reduced by introducing variable 'time of use' prices, for example, by charging higher power tariff rates at times of high demand, so that EV charging would tend to be done later at night, away from the large evening peak heat demand period. That could reduce stress on the power grid and also avoid the need that might emerge to build more power plants just to meet the extra evening peak demand in the winter. Even so, it was argued that it would be much easier to balance supply and demand if some form of energy storage was available. However, storing electricity is much harder and more expensive than storing gas, so the argument takes us back to the existing system, in which gas is stored and delivered by pipe to meet heat demand. Staying with that would mean that, rather than having to upgrade the power grid, the gas mains could still be used, and it also avoided the need to change heating devices in each home. Installing electric heat pumps would be very invasive and expensive.

Continuing to use gas thus seemed a much better idea, as long as it was possible to find a replacement for fossil gas, so as to cut emissions. The most obvious option was biogas, bio-methane produced by anaerobic digestion of biomass, for example, farm and municipal bio-wastes. However, the amount of biogas that was likely to be available from these sources was limited, and growing crops for biogas production faced land-use constraints. One study suggested that biogas production might be expanded to meet about 10% of domestic heating demand (Glover 2016). That may be pessimistic, for example, given the potential for using food wastes (with no new land-use implications), but more would be needed if biogas was also to be used for vehicles, as an alternative to the EV approach. Biogas is well suited to use in some types of vehicle, tractors, trucks and vans in particular, but producing enough biogas for that and beyond that might be very difficult.

However, one reason to be more optimistic was that there were some novel options for expanding *synthetic* gas production. As noted earlier (in Chap. 1), before the advent of North Sea gas, the so-called town gas was produced from coal. The UK could not go back to that (due to its high carbon content), but is possible to use fossil gas to make a new gaseous fuel, hydrogen, and capture and store the carbon dioxide gas produced during this process, making it low carbon. The so-called steam reformation process, done at high temperatures, is the main way hydrogen is produced at present globally. Adding CCS to it would make it nearly carbon neutral. The resultant low-carbon hydrogen could be injected into the gas main, mixed in with fossil gas or even delivered directly to users at 100%. That is the basis of the '100% hydrogen' H21 plan proposed for the city of Leeds (H21 2016). Hydrogen can also be used as a vehicle fuel, either directly in an engine, or to generate electricity for an EV power unit in an onboard fuel cell. Direct hydrogen-powered cars and buses, as well, increasingly, as fuel cell-powered EVs, are already in use in a range of UK locations, with the Department of Transport providing some funding to support the roll-out of hydrogen charging stations (DoT 2018b).

The fossil gas conversion approach to producing hydrogen does involve CCS, to make it low carbon, and that is as yet uncertain and likely to be expensive. An alternative is to make hydrogen by the electrolysis of water, using renewable electricity. That would be totally carbon free. However, this so-called Power to Gas (P2G) approach is at present more expensive than the currently dominant steam reformation of methane ('SMR') route, although adding CCS to the latter would reduce the cost advantage. Using surplus electricity from wind and PV solar, which would otherwise be dumped, should make P2G even more competitive. So, as the technology improves, it could be a winner (Beez 2017).

As we have seen, hydrogen can be used in a variety of ways, for heating, power generation and as transport fuel. If needed, the hydrogen could also be converted to methane gas by using captured carbon dioxide, or indeed to liquid synfuels like methanol, a so-called Power to Liquid approach (P2L). Both P2G and P2L are being developed quite widely in Germany (DENA 2018).

Note that, while, in theory, P2G (and P2L) could use bulk renewable electricity supplies, whenever available, more usually, use is made of *surplus* electricity from wind, solar and other renewables. There is likely to be plenty of this in future, since in a power system based on variable renewables, there would have to be a large renewable generation capacity so as to be able to meet most demand most of time, even when demand was high and renewable inputs low. That would mean however that, at times of low demand, there could be a large surplus produced, especially if renewable inputs were also high then. If this surplus could be used to make storable gas by P2G conversion, then that could be used to make electricity when power demand was high and/or renewable inputs were low. Renewable variability would thus in effect be turned from being problem into at least a partial solution. P2G would also reduce the need for curtailment of any surplus renewable electricity that was, or could be, generated. Curtailment (i.e. not taking available power) is economically wasteful, and, with large amounts of renewable input on the grid, could actually be more of a problem than shortfalls. P2G deals with that.

That view has yet to win the day, but P2G hydrogen production certainly offers some interesting answers to the problem of balancing variable renewables. So of course could other storage options, but hydrogen gas, like other gases, does have the attraction of being easier to store than electricity, and, it is claimed, P2G hydrogen is getting cheaper (Goodall 2017). So it might rival the cheapest and more commonly used storage option, pumped hydro storage, as well of course as batteries, although the latter are really only useful for short-term storage, whereas gas could be stored for long periods (Elliott 2017a).

A crucial point in this context is that balancing variable renewables needs a mix of short- and long-term responses. Short-term balancing measures for the frequent rapid variations and longer-term options for the occasional long lulls in renewable availability. Storage of P2G gas is one such longer-term option, possibly using underground cavern storage of bulk hydrogen. Note however that the power system does not run flat out, at full demand, all the time. The main issue would be meeting daily *peaks* in demand, and these could be reduced (delayed) by demand management systems, so reducing storage delivery requirements. So could power imports, assuming there was excess available from other countries when needed.

As we shall see, storage issues have featured centrally in the debate over the integration of variable renewables, and since heat storage is one option (heat being much easier to store in bulk long term than electricity), so has the production and use of green heat. Similarly, since EV batteries might offer some interactive storage capacity, when used in vehicle-to-grid (V2G) mode, the green transport issue has also, to some extent, become intertwined with the wider grid balancing issue. In addition, the development of new grid systems and grid management techniques has also become an important issue. A switch-over to a whole new system seemed to be both required and possible, enabling variable renewables to be used on a wide scale (Elliott 2016a).

There is as yet no agreed clear consensus on the way ahead, but the government's 2017 'Clean Growth Strategy', set within the context of its new overall Industrial Strategy, along with its *Smart Systems & Flexibility Plan*, offered ideas for the development of the new more flexible supply and demand system that would be needed for effective grid balancing. See Box 7.2.

Box 7.2 Industrial Strategy for 'Clean Growth' and Smart Grids

The government's 2017 White Paper on Industrial Strategy was mostly couched in very general policy terms, identifying four 'Grand Challenges'. It said we must 'put the UK at the forefront of the artificial intelligence and data revolution; maximise the advantages for UK industry from the global shift to clean growth; become a world leader in shaping the future of mobility; and harness the power of innovation to help meet the needs of an ageing society' (BEIS 2017a).

Some of this would involve specific sector deals, plans and programmes. For the energy sector, it said: 'We will continue to work closely with the nuclear and offshore wind industries to further drive down the costs of clean power, while building UK supply chains. We will also continue to explore the long-term options for clean heating and the many potential uses of low carbon hydrogen.' It also proposed a 'Prospering from the energy revolution' programme, 'to develop world-leading local smart energy systems that deliver cheaper and cleaner energy across power, heating & transport, while creating high value jobs and export capabilities. Our world-leading Smart Systems & Flexibility Plan will build on the rollout of smart meters to grow the markets for these systems & technologies in the UK.'

The Smart System and Flexibility plan, produced with Ofgem, was certainly quite ambitious on the power system side, laying out some key principles and aims. For example, it said: 'We want to help energy consumers gain control of their bills and how much energy they use. Smart technologies have a central role to play in this and can provide consumers with greater control and choice.' It also claimed that 'a smart system can deliver energy security at a lower cost than would otherwise be the case'. It also looked to wide gains: 'As a core part of the Industrial Strategy, [the plan] is an opportunity to increase productivity at home, and put the UK in a leading position to export smart energy technologies and services to the rest of the world' (BEIS 2017b).

On the transport side, the Industrial Strategy noted that the government 'was already investing over £600m in support to accelerate the transition to ultra low emission vehicles', and was expanding that with a further £270 million. Most of that was for EVs, and there was also funding for EV battery development, but biofuel and hydrogen options were also being supported.

The Industrial Strategy also of course looked to what could be done in manufacturing industry, with proposals for innovation in energy use, coupled with the switch-over to using renewables for energy supply. This increasingly can come from the power grid, but on-site wind power has already been used directly by Ford at Dagenham to power manufacturing, and there have been some interesting proposals in the primary materials production sector. For example, the use of local wind power has been proposed for the Lochaber Aluminium smelter near Fort William in Scotland (Musaddique 2017), and it has also been proposed to use biogas, from an anaerobic digester fed with food and other waste from a nearby waste recycling centre, for a forge in Sheffield (REM 2017). In addition, GFG Alliance's 'Greensteel strategy' aims to cut the amount of raw steel imported to the UK, by dramatically increasing the amount of scrap steel which is recycled. It planned to use electric arc furnaces, part-powered by renewable energy, to melt scrap steel so it can be reused, a process which is more environmentally friendly than primary steel-making in a blast furnace powered by coal. It was claimed that 'Greensteel, made using renewable energy, has only one tenth of the carbon footprint of blast furnace production' (Tovey 2017).

As can be seen from Box 7.2, in addition to smart grid ideas, there are some interesting ideas for manufacturing and others sectors, as well as ambitious plans for the longer term. However, the government's main practical system-level proposal in the energy supply sector in the short term was to build on the smart meter programme. We will be looking at that later, but it is relatively limited and also progressing slowly, and it is likely to take some time for the more advanced grid balancing and smart grid programme to emerge fully, as part of an integrated system-wide change.

In what follows, I trace the way the grid balancing and smart grid issues unfolded, running from the early days to the present, with examples of UK government responses, before looking at responses to the linked 'green heat' and energy efficiency issues, and then at what might happen next to bring it all together into an integrated and balanced system.

7.2 Grid Balancing

Looking back to the earlier stage of renewable deployment, from the government's point of view, there was at first no great need to take the balancing issue too seriously. With, initially, in the 1990s and early 2000s, relatively small amounts of renewable capacity on the UK grid, its variable availability did not matter too much. The small variations were lost in the much large variations in (other) supply and in demand. These were dealt with by a range of balancing measures, mainly by varying the output of some of the fossil fuelled plants or topping up with power from pumped hydro storage to meet any temporary shortfalls. A few plants were also kept in reserve for longer periods of backup. Import of power from abroad was also an option.

It was claimed (e.g. in the early days, by the CEGB) that as renewable grid inputs expanded, these measures, possibly expanded slightly, could cope with the small increasing variability, at least up to around a 10–20% renewable contribution, and this view was fairly widely accepted (Grubb 1991). Subsequent studies and experience in countries like Denmark, where quite high shares of renewable were being introduced, seem to confirm this view (Milborrow 2007; Boyle 2008), although it was noted that as the renewables share went higher, balancing would become harder. Extra balancing measures would be required. That would add to the cost, but the adjustments needed were relatively straight forward, just extensions of what was already done.

However, this relatively sanguine view was not accepted by all. Some critics of renewables insisted that, inevitably, they would need major extra backup (some said each plant would need 100% backup), since they could not run all the time. For example, a study funded by the arguably oddly named Renewable Energy Foundation concluded that wind power could not be relied on since it varied too much around the UK, with long lulls at times (Oswald et al. 2008). A response by researchers at Imperial College London noted that, although the title of the paper was 'Will British weather provide reliable electricity?', the substance of the paper was about the analysis of the impact of variable wind on *individual* generators, which they said was a different issue. What mattered was the impact on the system as whole, as had been the focus of earlier work by the Imperial/UKERC team, reviewing a range of studies which had used a 'time series' statistical approach, based on actual plant data, rather than just extrapolations from general Met Office wind data (Gross et al. 2006).

The system-level approach highlights the fact that what mattered for system reliability was the net balance of supply and demand over the whole system, not the output from individual plants. All power plants could fail at times, but no one suggested that dedicated backup plants be provided for each individual one. The system as whole provided backup.

Even so, the system did have to provide backup, and some critics argued that, by having to ramp up gas turbines regularly, more carbon dioxide would be produced than was avoided by the use of variable renewable energy plants, when available. This argument was relatively easily dealt with. Yes, there would be a small amount of extra carbon dioxide produced by ramping up the gas plants when renewables were offline for a while, but that would be far less that was saved by the fact that the renewable plants would run most of the time. Essentially they would act as net fuel (and carbon) savers.

A more nuanced point was then raised. Since the gas plants would mostly not be running at full capacity, they would be less efficient and would produce more CO_2/kWh of output than they would otherwise. The counter argument was that the CO_2 output would still be much less than that avoided by the renewable plants overall. A 2008 report from a House of Lords Select Committee noted: 'The need to part-load conventional plant to balance the fluctuations in wind output does not have a significant impact on the net carbon savings from wind generation' (Lords 2008).

The critics however were not convinced. They pointed to low wind plant *load factors* (sometimes also called *capacity factors*), these being a measure of the actual output compared to what could in theory be produced if the plant could be run flat out all the time. Typically, they were around 20% averaged over the year in the UK at that point, which meant some were much lower, some (on good windy sites) much higher. But the critics sometimes used load factor figures to suggest that the plants would only run '20% of the time', or less in some locations. So they were inefficient and of no use. In reality, even if they have low annual load factors, most wind turbines run 70–80% of the time at *some* level of output. The availability of this output and its correlation with demand, and with other supplies, is what mattered in terms of system-level reliability. Obviously, getting to higher load factors was important, since you then got more output for your investment in the plant. Much higher load factors have now been achieved, up to 50% or more, especially for the new larger offshore wind turbines (8 MW and more), run at sites where the wind regime is more reliable. So we now hear less of the simple arguments about the operational unviability of wind.

We have however heard more about the cost of balancing variable renewables. Some put it very high, but also included the earning losses that would be incurred by gas plants that were finding it hard to compete in the market at times when cheaper power was available from renewables. This had become a significant issue in Germany in the late 2000s, where PV was pushing gas plants out of some peak, high value, markets, so that they were uneconomic. Some saw this so-called profile cost as perhaps doubling the overall cost of renewables (Ueckerdt et al. 2013). But was it fair to treat the commercial losses the rivals incurred as a cost of renewables? The balancing cost should, arguably, just be the extra cost incurred by the system having to supply power when renewables were not available.

At low levels of renewable input, the overall operational balancing cost would be small or even negative. The gas plants already existed, built and paid for, on the grid, and just had to run more often at full power, the rest of the time they could be ramped down, saving money (and emissions) by not burning fuel. Nevertheless, with higher renewable contributions, more gas plants, or other backup options, might have to be made available. Depending on which options were used, and the scale of the renewable input, a review by the UKERC of the studies that had been done so far concluded that balancing might add at most (at high renewable input levels) around £15/MWh to the cost of power, about

10% of then current offshore wind-power costs. If the so-called profile costs were also included (what UKERC called reserve costs), with wind and solar supplying 30% of electricity demand, there might also be an extra £5/MWh imposed, that is, from earning losses incurred by rival plants mostly not being used, but kept in reserve. At a 50% wind and solar penetration, this cost could rise to between £15 and £44/MWh, but with the top end reflecting a very inflexible electricity system. The point is that the cost depended on how the system was configured. Keeping large plants in reserve was not necessarily the best option. For example, if peak demands could be reduced, it all got easier, and cheaper. It was the same if storage was available, or imports from abroad. What was needed was a flexible supply and demand system, with the right mix of balancing systems. Indeed, the UKERC study noted that its data showed that, without this, balancing costs would rise dramatically, but with it, the extra cost could be relatively low (Heptonstall et al. 2017).

Warnings like that were likely to be noticed by the government, which otherwise had been more concerned about having enough supply capacity to 'keep the light on'. On that basis, if there were going to be occasional shortfalls due to renewables, then the simple response was to just add more capacity or backup reserves. The more nuanced message suggested that there might be cheaper routes. Gas turbines are relatively low cost, so it might make sense to add a few more for occasional use, and also possibly cheap diesel standby sets. They would not be used often, so their emissions would not matter too much. However, a whole new fleet of gas plants was not needed: it would arguably be more efficient, and possible cheaper and certainly cleaner, to develop a flexible demand system, with smart demand management for peak-shifting, and storage, toped up if necessary with imports, to meet residual peaks and/or shortfalls.

The storage idea, the use of batteries in particular, caught the attention of government first. It seemed obvious that this was what was required to deal with variable renewables. Indeed, it became almost an article of faith that, without storage, renewables were of little use. Thus, in evidence to the Energy and Climate Change Select Committee in July 2015, Amber Rudd, then Secretary of State for Energy, said 'until we get storage right, renewables are unreliable' (Rudd 2015).

There certainly has been much emphasis put on storage, and, in addition to new types of more efficient batteries (Lithium Ion batteries are currently the leader), there are many new options emerging, including liquid air storage systems and advanced flow batteries (Elliott 2017a). The UK has a presence in the field, although a veil should perhaps be drawn over one of its early attempts to develop an advanced flow battery, the Regenesys regenerative fuel cell project. Much trumpeted at the time (the early 2000s), its development was stalled and then abandoned (MPS 2003). However, technology moves on, and the storage field is currently booming, including Lithium Ion batteries for home use, with PV solar, although, as Box 7.3 suggests, the cost can be high.

Box 7.3 Domestic Battery Costs and Benefits

Telsa's 13.5 kWh rated Lithium Ion Powerwall domestic battery unit currently retails in the UK at nearly £6000, including an inverter (Tesla 2018). Assuming a domestic user could charge that fully during the day from a typical 2 kW PV array on their roof, it would power a standard 2 kW heater, or something similar, overnight (i.e. in this case, for 6.7 h). If they wanted more power than that at night (e.g. to charge an EV and/or run a washing machine for a while overnight), they would need more batteries and more PV. Moreover, since on some days there would not be much output from the PV arrays, they would need even more batteries and probably more PV arrays, multiple units, to cover, say, a week of dull skies. Clearly, at current prices, that is unviable for most home applications, and most domestic PV users will have to rely on grid imports to give full power cover, as is currently the norm for most users. Major further PV and battery prices falls might allow more capacity to be installed and so reduce the need for grid top-ups, but in most locations and situations, there are likely to be diminishing returns from trying to go fully 'off-grid'. Indeed, as we shall see later, two-way grid interactions (via import and export links) may become increasingly important as we move to integrated smartx grid systems, with distributed generation and storage playing helpful grid balancing roles (Elliott 2017a).

Clearly, storage can be expensive, especially if we are talking about longer-term storage, for more than a few hours. Some types of larger-scale battery storage systems are now cheaper, including flow batteries, and some are being deployed for short-term utility-level grid balancing, but that is only part of the answer to balancing, short and long term.

Certainly, dealing with balancing options one by one in isolation can be suboptimal, as several academic studies and agency reports indicated (Elliott 2016b). As IRENA put it, 'energy storage is only one of many options to increase system flexibility' (IRENA 2015), and it was flexibility that was the key thing. That meant the use of a mix of balancing options, including demand management. For example, time-of-use tariffs could be used to incentive consumers to shift their power usages to times when demand is lower and/or renewable supplies more available. Unlike storage, that flexible approach involves no extra capital expenditure, apart possibly from smart meter switching software to automate the process.

This message about the need for a flexible system was put in a more positive and (politically) influential form by the National Infrastructure Commission that the government had set up in 2015 to oversee major project development plans, including for energy. In a report on 'Smart Power', it said that an integrated flexible supply and demand management system, with smart grids, storage and also grid interconnector imports, could save the UK £8 billion p.a. by 2030 (NIC 2017). The point was that the new system could be more cost-effective than the existing one, matching supply and demand more effectively locally and maybe also more widely, and avoiding energy waste. It might also perhaps be added that, by enabling variable renewables to be used, it would also cut overall energy costs, since that would reduce the use of expensive fossil and nuclear energy.

The result was that, following the production of the 'Clean Growth Strategy' in 2018, the government-funded research programmes on integrated smart grids and system management, as well as on storage, including hydrogen and heat-based systems. For example, the government-backed Innovate UK offered up to £41.5 million for smart energy projects that link supply, storage and demand in heating, power and transport. The funding was part of a £102 million 'Industrial Strategy Challenge Fund 'prospering from the energy revolution' (Innovate UK 2018). In parallel, the government announced a £20 million Hydrogen Supply initiative, as flagship 'HyNet' project, and also offered support for significant heat network development, in a £320 million programme (BEIS 2016a).

We will be looking at heat networks more later (Sect. 7.6), but they are of special interests here since they could play a role electricity balancing. For example, heat networks can be supplied by CHP plants, in which the heat energy that is wasted in conventional plants is captured. CHP plants can be used flexibly, in conjunction with heat stores, to balance the variable output from renewables, by varying the ratio of their heat to power output and feeding any surplus heat to the heat store. If there is too much green power on the grid, the CHP plant can produce mostly heat. If demand for that is low, it can be stored. If green power availability is low, the proportion of CHP plant power output can be raised, and if there is still demand for heat, it can be drawn from the heat store.

Although CHP does need a nearby heat load to serve, in the power sector, it can be a flexible and valuable transitional option, complimenting renewables, and capable of reducing emissions/kWh significantly, or entirely, if run on biomass. CHP can also be used in the industrial context, meeting power and heat demand directly and reducing emissions.

The UK has not developed CHP very significantly outside of industrial projects. It still only has under 6 GW, 70% of it is industrial and mostly gas fired, with no growth since 2010 (DUKES 2018). By contrast, CHP is quite widely used for community heating (amongst other things) elsewhere in the EU, linked to district heating networks. Increasingly, these networks are being fed with heat from biomass sources (directly or via CHP), with in some cases large heat stores being added and being topped up by solar heat. In the UK, by contrast, at present, heat networks are rare and only supply around 1% of building heat demand. However, in the government's 2017 'Clean Growth Strategy' scenarios, heat networks were projected to meet 17% of heat demand in homes and up to 24% of heat demand in industrial and public sector buildings by 2050 (BEIS 2017c).

With high-efficiency low-carbon CHP plant being one option for supplying the heat, cost can be relatively low, but there have been concerns about how consumers are charged. Heat is less easy to measure than power. There were also worries about reliability. A Competition and Markets Authority study set out to establish whether heat network customers were getting a good deal in areas such as price, quality and service levels. It found that, for many, heat networks offered prices which were the same or lower than those for electricity or gas tariffs and had similar levels of customer service. But some users, mainly those living in privately owned or rented properties, paid more for their heat via a heat network and, across the board, heat network users did not get the same levels of protection as gas and electricity customers. So there can be some management issues (CMA 2018).

That will also be true of full 'smart grid' systems, with 'demand side management' (DSM) or 'demand side response' (DSR), integrated with the use of storage, especially if heat storage was to become more important for grid balancing. That would all operate at the regional and national level, presumably with overall management by National Grid. But in addition, power imports from the EU, via extra grid interconnectors, could possibly play a major role, balancing out local variations in supply and demand over a much wider geographical area. See Box 7.4. That would open up even more management issues. Clearly all these new grid balancing and demand management systems can interact, and seeking integrated and optimal approaches can be difficult: it is early days.

Box 7.4 Interconnectors

The UK National Infrastructure Commission (NIC) said that 'Interconnectors offer a number of benefits to the UK and are a key source of flexibility to the electricity system. They are one of the few existing technologies that can shift large volumes of electricity from where it isn't needed to where it is. By doing this they have the potential to lower prices for consumers, improve the investment case for power stations, help us meet our carbon targets at a lower cost and improve security of supply. Interconnectors can be thought of as both additional generation capacity and power exporters. This is because they allow electricity to be imported at times of peak demand but are also able to sell electricity abroad when we have more than we need.' So it wanted more than the 4 GW the UK then had, and maybe more than the extra 7–9 GW planned (NIC 2017).

There can certainly be some benefits from interconnectors. Poyry consultants noted that 'There is strong agreement that the socio-economic welfare value of interconnectors comes from their ability to improve the efficiency of outcomes in the electricity system, lowering the cost of meeting demand and of achieving other policy objectives such as improving security of supply and enabling more efficient renewables integration.' They added that 'the overwhelming majority of literature reviewed concluded that additional interconnection beyond current levels is likely to bring significant benefits to the UK' (Poyry 2016).

Poyry said the overall costs and benefits of links were 'overwhelmingly positive' for the UK up to 9–11 GW, though beyond that it was less clear. Moreover, while it can be argued that an EU-wide system would have overall benefits for the EU, that might not always be true for individual EU countries. That depended on market conditions. For example, there may not always be demand for UK surpluses or available supplies at reasonable costs when the UK had a shortfall, and some of what might be available would not be 'green' power.

However, it is early days so far, and new trading relationships and markets would no doubt emerge, while the latter problem should lessen since most countries in the EU are aiming for high shares of renewables. Some countries also have large hydro reservoirs, some of which can be used for pumped storage. A High Voltage Direct Current (HVDC) 'supergrid' network, with much lower energy losses than conventional AC grid links, could allow other countries to access that storage capacity, as well as linking up renewable generation capacity across a wide area, to aid balancing (Elliott 2012). Indeed, some look to extend the EU supergrid concept to include North Africa, so that some of its potentially vast solar resource could be tapped, creating new opportunities for trade and balancing (FISES 2016).

While as noted above, the UK government began to support research on specific hardware projects (e.g. in relation to storage) and software (DSM/DSR), and on system integration generally, in terms of how the overall balancing system might evolve, it decided to leave that mainly to the market, establishing a capacity market, in the belief that market competition would select the least-cost options. As we shall see, this has not necessarily yielded an optimal mix, in part, arguably, since the government seems to have tried to use the capacity market as way to provide extra support for nuclear and fossil options, rather than for new flexible grid balancing systems. However, it is also true that the latter were, initially at least, more expensive, so the market approach did not favour them.

7.3 Capacity Market

The UK government's approach to the capacity market was essentially to use it to try to 'keep the lights on' at lowest cost, via competitive auctions. The initial concept was to support supply capacity that could be used to meet shortfalls and maintain security of supply. The government already had the option of calling on some capacity kept in reserve, including some mothballed plants, and there was STOR, the Short Term Operating Reserve system, in place to provide short-term balancing. But the capacity market was meant to be something much larger and more comprehensive, able to also deal with significant longer-term short falls, whether due to plant or grid failures, or renewable supply variations and lulls.

The first four-year ahead Capacity Market auction, in December 2014, led to around 49 GW of capacity being contracted, out of about 65 GW of submissions, for reserve/backup duties in 2018–2019, if required. Most of it was existing or refurbished fossil-fired capacity, along with near 8 GW of existing nuclear. See Table 7.1. Note that 49 GW was a large proportion of the non-renewable capacity (fossil and nuclear) that was on the grid, plus a contribution from (non-variable) hydro. So almost everything that was contracted to be available if needed was existing plant. For this, they would receive a payment, the cost being passed on to consumers—another levy.

The specific allocations raised some hackles. Professor Catherine Mitchell at Exeter University said that it was 'supporting the status quo', with the vast majority of payments going to large, well-established generators: 'They have done a great job in persuading Government that unless customers pay this extra amount it would be uneconomic for them to keep the power plants running and the lights will go out. All of this money would have been far better used to develop a sensible demand side response market' (Mitchell 2014).

Table 7.1 Capacity market first auction round

Capacity (MW)		(as %)	Number	(as %)
CCGT	22,258.6	45.19	47	15.36
CHP & autogeneration	4235.1	8.60	36	11.76
Coal/Biomass	9232.2	18.74	29	9.48
DSR (demand side response)	174.2	0.35	15	4.90
Hydro	682.2	1.38	29	9.48
Nuclear	7876.1	15.99	16	5.23
OCGT and reciprocating engines	2101.2	4.27	121	39.54
Storage	2699.4	5.48	13	4.25

National Grid (2014)

Note that most contracts were for one year, but some were for longer: see Box 7.5

As it was, only a very small amount (0.35% of the total) went to DSR, but 2.7 GW did go to storage, though much of that was existing pumped hydro. The success of gas plants, large and small (around 25 GW in all), was not surprising, but the fact that coal (including biomass co-fired plants) also got shares was seen as worrying, given that their use would expand emissions. Dave Jones, an analyst at Environmental group Sandbag, said that 'The capacity market looks more like a subsidy scheme to keep heavy polluters online, rather than as a mechanism to encourage new investment – only five per cent of auction revenues will go to new investment. The capacity mechanism is actually slowing decarbonisation of the UK power sector' (Jones 2014).

Moreover, there were concerns about longer-term capacity market shortfalls. *The Telegraph* pointed out that only one large new gas plant was contracted for and felt that this was odd given the alleged yawning 'energy gap' due to expected plant closures. It reported: 'critics have suggested the policy is turning out to be poor value for money as it will hand vast subsidies to old nuclear plants that would have kept running anyway, and to old coal plants that are simultaneously being subjected to environmental taxes designed to force their closure' (Gosden 2014a).

In similar vein, *Carbon Brief* complained that the capacity market 'awards payments to plants that would have been open anyway – in the first auction around a third of the plants which won contracts signaled they would have stayed open with no or very little payments'. It added 'The market is also focused on the needs of large fossil fuel and nuclear plants rather than new technologies which can reduce costs to bill payers by shifting demand. Such technologies can reduce the need to build expensive new power stations and the amount of time that the most expensive stations need to run' (Orme 2014).

So it seems no one liked the capacity market, apart from the winners and the government. However, the auction prices that emerged were lower than expected, at £19.40/kW (totalling £990 million), down from the initial 'reserve auction price' of £75/kW. That was good news for consumers, but, even so, it still meant an extra charge to them. When the capacity contracts in this first round went live, from 2018 onwards, they were expected to put around £11 p.a. on consumer's bills. That prospect was not something the Energy and Climate Change Select Committee took kindly to. Its chair, Conservative MP Tim Yeo, said that it meant 'every consumer in the country […] subsidising spare electricity generating capacity that may only be used for a few hours each year. But smart technology has now made it possible to reduce unnecessary electricity demand at peak times, thereby reducing the number of polluting power stations that need to be switched on. This could mean we can reduce the total electricity generating capacity that has to be maintained in future, bringing down costs for consumers while enabling us to reduce consumption of fossil fuels. Yet this promising new demand-side response technology has been disadvantaged in the auctions under the government's capacity market – meaning costs and emissions could be higher than necessary' (Yeo 2015).

Views on what should be done next were mixed. Given expected plant closures and the increase in variable renewables, it made financial sense to back gas plants, including existing plants. New ones might be needed, but upgrading existing one was cheaper, and both options were cheaper than storage. However, demand management was surely going to be the cheapest option of all, long term. Interconnectors might also have a role. As with DSM/DSR, that avoided having to build any new UK energy plants. Certainly, few thought that large inflexible nuclear plants could help with balancing much, so it was odd that they were included in the capacity market, and, arguably, even odder that more even larger nuclear plants were planned, an issue we will be coming back to later (Box 7.12).

There were some suggestions for improvements to the Capacity Market system. For example, *Carbon Brief* thought it might be helpful to have separate auctions for new and existing capacity. The Lords Science and Technology Select Committee, in a detailed commentary, suggested that electricity storage might usefully be placed under the Contracts for Difference regime rather than in the Capacity Market. See Box 7.5 for their recommendations and the government's responses.

Box 7.5 Capacity Market—Lords Select Committee Review and Responses

The House of Lords Science and Technology Select Committee report on 'The Resilience of the Electricity System' made a number of recommendations, many of them related to the Capacity Market (CM), to which the government then responded in 2015 (DECC 2015a).

The Lords wanted the Government to ensure that DSR 'is not disadvantaged in the Capacity Market relative to generation' and recommended that 'the length of DSR contracts in the Capacity Market should be brought into line with generation'. In its reply, the government said the 'Capacity Market was developed with the sector and includes specific features to promote the participation of DSR resources. However, to encourage a mix of technologies and promote competitive auctions and therefore value for money to consumers, the Capacity Market auction is technology neutral. The auction mechanism is driven by cost-effectiveness to determine the most suitable providers and the introduction of specific targets in the Capacity Market would lead to less competitive auctions increasing costs to consumers.' It added 'One year capacity agreements are the default position in the Capacity Market with fifteen-year capacity agreements being available only to new build generation which requires high up-front capital investment. Current evidence suggests that DSR is a relatively low-cost solution and should therefore be able to compete effectively on the basis of one-year agreements.'

The Lords wondered whether electricity storage should be placed under the Contracts for Difference regime rather than in the Capacity Market. The government noted 'that other countries are approaching energy storage in different ways. A few countries, such as Germany & Japan, which have more substantial or more urgent system balancing challenges, are using operating incentives to support energy storage deployment. We do not want to introduce targets which could lead to the deployment of one particular balancing mechanism where other solutions may offer a more cost-effective or lower-carbon solution to deal with intermittency of supply or other balancing problems.'

On interconnectors, the Lords said there was 'a worrying lack of clarity about what options exist if a number of interconnected countries experience system stress simultaneously'. The government said 'According to historical analysis of system stress events between GB and its interconnected countries conducted by Poyry for Ofgem in 2013, the expected correlation with interconnected countries is likely to be low, despite some coincidence in times of peak demand. There are good reasons for expecting the future contribution of interconnectors at times of system stress to increase relative to this analysis.' e.g. since 'GB wholesale prices are becoming more cost-reflective in times of system stress', and while 'market arrangements mean flows have not always followed prices', that should be addressed 'through market coupling reforms'.

It was noted that DECC had also commissioned a *Whole System Impacts of Electricity Generation Technologies* model, looking at the impacts of electricity generation technologies, including intermittent renewables, system costs and wider system balancing issues.

The debate continued, but the basic issues seemed clear. So were the risks. The overall approach to the capacity market was said to be 'technology neutral', so in theory all options should have been eligible, although clearly it all depended on costs. Some feared the auctions would yield a preponderance of cheaper but less environmentally appropriate options, for example, there had been media stories about hundreds of diesel generators being used. As can be seen, something near that did emerge and also in the next auction round, with around 800 MW of diesel sets contracted, along with 2–3 GW of small reciprocating gas-fired plant.

Overall, the second capacity market auction round in December 2015 (for 2019–2020) was slightly less costly than the first round, clearing at £18/kW, although only securing 46.4 GW in total. Around 7.5 GW of existing nuclear was contracted, along with 4.4 GW of old coal plant, as well as 1.8 GW of existing interconnector capacity. Storage was more or less unchanged at 2.6 GW (including old pumped hydro), but DSR got 456 MW of contracts, significantly up on the 174 MW in Round 1. Nevertheless, as before, there was a heavy bias towards gas, including 24.7 GW of existing gas plants and 801 MW of new gas plant (DECC 2015b).

Interestingly, a sense of urgency then seemed to emerge, with, in addition to the usual four-year ahead round (T-4) in 2016 (yielding 52.4 GW for 2020–2021), a special one-year ahead 'transitional arrangements auction' (T-A), held just for DSR in 2016, resulting in 800 MW of contracts for 2016–2017. Then, in 2017, there was an 'early auction' brought forward by a year, with the delivery date for the 50.4 GW contracted being 2017–2018, in response to emerging security of supply concerns. Also in 2017, there was another transitional auction (T-A), just for DSR, for 2017–2018, this one yielding only 300 MW. See Table 7.2.

With the pace evidently speeding up, as Table 7.2 shows, in 2018, there was a one-year ahead (T-1) auction for the winter of 2018–2019, followed by a T-4 four-year ahead auction for 2021–2022. BEIS described the one-year ahead T-1 auctions as providing an 'opportunity to correct for changes in required capacity', that is, fine tuning and adjustments of the T-4 targets (BEIS 2018).

As in all the previous rounds, in the 2018 rounds, gas still dominated, so, apart from the special DSR rounds, the pattern was not changing much, although DSR did a little better, as did interconnectors, while coal was more or less excluded. See Box 7.6.

Box 7.6 The 2018 Capacity Market Auctions

In the 2018 T-1 auctions, gas projects won 4.37 GW worth of contracts, 76% of procured capacity, and in the T-4 auction, gas got a massive 23 GW of contracts. The 2018 T-4 also saw nearly 8 GW of nuclear again getting contracts. By contrast, there was only 89 MW of battery storage capacity procured in the 2018 T-1, along with 3 MW of compressed air storage, making up less than 2% of the capacity procured, while DSR made up a further 7.5%, with 443 MW of DSR capacity procured. In the 2018 T-4, battery storage only won 153 MW, DSR did better at 1.2 GW and interconnectors went up to 4.5 GW, with 2.15 GW of new links. However, apart from that, there was one other significant change: coal was decimated. About 2.8 GW of coal plants exited the T-1 auction, with no contracts (Coyne 2018; Energy Manager 2018).

The allocation pattern may not have shifted much over the various rounds, but the competitive auction approach did push cost down for the main options (leaving aside the special DSR auctions), as can be seen from Table 7.2. Thus, the 2018 T-1 cleared at £6/kW per year, while T-4 had a clearing price of £8.40/kW p.a., these being around a third of the £18–22 price in earlier full rounds (Stoker 2018).

However, these savings came at a cost. In the 2018 T-4 round, only 762 MW was new build, out of the 50 GW contracted in all. Basically, once again, existing plants were just being subsidised. Only one new Combined Cycle gas project won out in T-4. That did seem short sighted (Vaughan 2018a). With old plants likely to retire, at some point there was a risk that there might not be enough gas-fired capacity for grid balancing, especially since renewables were getting so cheap: there was no incentive to build new gas plants, and old plants could not compete in

Table 7.2 Capacity Market overall auction results

Auction	Delivery year	Capacity secured (GW)	Clearing price (£/kW)
Dec. 2014 T-4	2018–2019	49.3	19.40
Dec. 2015 T-4	2019–2020	46.4	18.00
Jan. 2016 T-A	2016–2017	0.8	27.50
Dec. 2016 T-4	2020–2021	52.4	22.50
Jan. 2017 Early Auction	2017–2018	54.4	6.95
Mar. 2017 T-A	2017–2018	0.3	45.00
Jan. 2018 T-1	2018–2019	5.8	6.00
Feb. 2018 T-4	2021–2022	50.4	8.40

BEIS (2018)

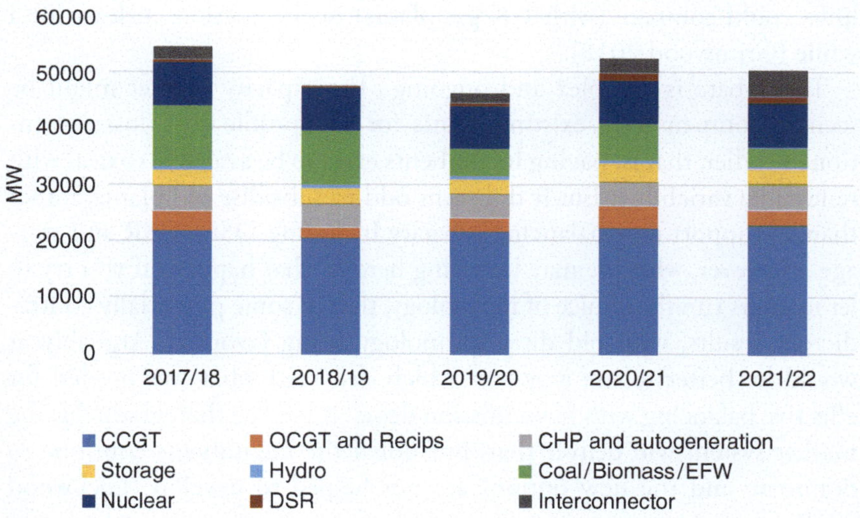

Fig. 7.3 Capacity market auction contracts by technology. (BEIS 2018)

the new market. Figure 7.3 summarises all the allocations by technology in the auctions so far.

At one time, it looked like the capacity market *was* going to be used to subsidise new gas plants, since it was thought that they were vital for the future, but as we have seen, the low auction prices have denied them much access, and in any case BEIS projections now had gas use falling (see Fig. 6.5). Optimistically, some have hoped that, in that case, storage and DSR could partly take over as balancing options, but as noted above, so far very little storage has been contracted and not much more DSR (1.2 GW in the 2018 T-4).

As also noted above, grid interconnectors did better in the 2018 T-4 round, at 4.4 GW, about half of that being new links. On that basis, some said that, with only 762 MW of new-build gas plants, interconnectors were taking the lead and should be relied on more. The counter view was that, although they did produce emissions, gas plants were good for flexible balancing. While interconnector links did offer balancing capacity (see Box 7.4), they might not be as reliable as gas plants or other balancing systems *based in the UK* (Vaughan 2018b). Moreover, some of the links might be to old coal plants overseas, that possibly being why these links could compete (with UK gas plants) in the auction, at least for a while (Grimwood 2018).

The debate is complex and ongoing. The capacity market might be seen as compensating existing plants for the 'profile cost' losses mentioned earlier, that is, paying incumbents extra to be available to deal with renewable variability. But it did seem odd to subsidise old plants, rather than to support new balancing capacity including DSM/DSR and storage. However, what we may be seeing here is what happens if you try to let markets run the choice of technology, that is, some potentially contradictory results, with old dirty technology being favoured. Arguably, it would be better to use a system which delivered what was needed for effective balancing with low-emission debts. It may be that, gradually, the market system will deliver that, but not if the incumbents continue to dominate and the new options are not helped to develop (Lockwood 2018).

The wider message seems to be that, if you want new technology fast, including systems for balancing, you do have to provide targeted market subsidies or direct development funding. That is not surprising. Trying to cut corners by repurposing old conventional plants can be less costly, but counter-productive, it just keeps them going. Put simply, you have to invest to save, that is, investing in what may seem expensive now will hopefully lead to reductions in cost and long-term overall major savings.

The Government's Smart Systems & Flexibility Plan noted that one estimate of the long-term benefit of a smart energy system was £17–40 billion to 2050, due to 'avoided or deferred network reinforcements, avoided generation build, avoided curtailment of low carbon generation, and better operation of the system' (Sanders et al. 2016). That savings estimate of course covers quite a wide range, and there are also significant uncertainties on the investment cost side, so the overall cost-benefit ratio, while hopefully clearly positive, is not yet confirmed, though it does look promising.

The Capacity Market is still relatively new. It will take time to see if it, and the smart grid management system that it will hopefully support, is effective as more renewables come on line. However, in parallel, as something of a precursor to smart grid management, the UK is in the midst of rolling out smart meters. That might be seen as a test run for the larger smart grid system. So how has it gone?

7.4 Smart Meters

The £11 billion Smart Metering Implementation Programme (SMIP), initially proposed by DECC in 2009, aimed to install smart meters in all the UK's 26 million homes, as well as other premises, by 2020, eventually involving 53 million meters in all (DECC 2013a). It was claimed that their use would help reduce energy demand and save both users' and suppliers' money: estimates ranged around £6 billion in total over the programme life.

Smart meters send energy suppliers data on consumer energy usage automatically and wirelessly each month, so consumers get accurate bills instead of estimated ones. That also allows consumers to monitor their energy use in real time. Some versions provide a separate Smart Energy Display (SED) for consumers to use. The expectation was that this system would make consumers aware of their energy use, so that they might then reduce it, and their energy bills. Suppliers would benefit by not having to employ meter readers and by having full data on their customers' energy use, which could aid system planning and cut costs.

Unfortunately, all has not gone well with the smart meter roll-out. As with any advanced IT system, initial bugs and delays were to be expected, but the smart meter programme seems to have attracted more than its fair share and this has led to much adverse media coverage. Some of it was initial and generic (Rose 2014), some was on specific post-installation glitches like faulty billing (Shannon 2015; Meadows 2017; Morley 2017), while others focused on the cost, which would eventually be passed through to bills (Gosden 2013; ITV 2017). The response from some energy professionals was sometimes also very critical (Hunn 2013; Buchanan 2015; Means 2017). The Consumer Association's *Which* was also very pessimistic about likely roll-out rates (Pratt 2018a).

Nevertheless, the government stoutly defended the programme, despite its problems (Bourne 2017). Survey work had apparently found that, once installed, the meters were welcomed by most users (Whitlock 2017; BEIS 2017d). However, a study of the smart grid programme by researchers at Sussex University pointed to lack of consumer engagement, insufficient information and inadequate attention to vulnerability, which it said were key reasons for the slow roll-out, this despite the £100 m marketing campaign. It claimed that there had been consumer apathy and confusion, especially in the case of vulnerable people, with uncertainties about the benefits remaining. It noted that, at one time, it was said that the programme might cut energy use by 5–15%. But now 1–3% was seen as more realistic, with, it seems, some targets abandoned (Sovocool et al. 2017).

Clearly, there were issues with the scheme, and some of them relate to the inflexible technology that has been chosen. It will probably need to be replaced if the UK wants to develop a full smart grid system: See Box 7.7.

Box 7.7 The UK Smart Meter Experiment

The UK smart meter scheme started to lift off at trial stage in 2012 (HMG 2013), but it took time to get going fully, with technical and administration problems and delays. By the time of the initially planned 2014 full-scale launch, according to DECC, there were 'more than 400,000 smart meters in homes and over 490,000 smart and advanced meters in non-domestic premises' (DECC 2014a). By 2016, 2.3 million smart meters had reportedly been installed in UK homes and 4.9 million overall, and 8.6 million overall by 2017 (BEIS 2017e).

There had evidently been problems with the first-generation SMETS1 meters. By December 2016, some 330,000 of these meters were reportedly just running in 'dumb' mode, not operating as smart meters, and by March 2017, that figure had risen to 460,000 (Renew 2018). A major issue was that, it seems, the meters would not work if consumers switched suppliers. Professor Dieter Helm said that showed the fatal flaw in the way the roll-out was set up. Responsibility for it should have been given to the National Grid, not to the individual energy supply companies, who, naturally enough, each promoted their own systems. So changing suppliers was hard (Helm 2018).

That issue is evidently being at least partly remedied with an update, the SMETS2, which is more flexible. Nevertheless, none of the meters as yet are designed to switch to the cheapest tariff automatically, responding interactively to *pricing* signals, and rival offers. Much less can they oversee peer-to-peer 'prosumer' power trading, or electric 'vehicle to grid' power exchanges. That sort of facility may be possible later, but would need new technology (Ambrose 2017; Edie 2017).

With the whole energy system undergoing major changes, it did seem odd just to install very simple readout-only meters. The advent of peer-to-peer trading of power from domestic and independent PV solar systems, and electronic payment systems like Solar Coin (Gurzu 2017), indicates the need for more flexible and smarter systems. There certainly are some interesting projects going on elsewhere in the EU (Sonnen 2018).

However, there can be problems. Electronic credit exchange systems open up new issues, as indicated by the recent bitcoin saga, with allegedly very heavy energy use being one issue for the Block Chain system (Cuthbertson 2017; Di Christopher 2017; Auffhammer and Wolfram 2018). One recent study suggested that 'the processes involved in a single Bitcoin transaction could provide electricity to a British home for a month' (Truby 2018). Nevertheless, Block chain systems may aid peer-to-peer trading of 'prosumer' self-generated power, and there has been a test project trading surplus power in the UK, with more planned (Richardson 2018; Reuters 2018).

This may be one way smart grids develop, locally, from the bottom up, but it seems more likely that if the UK is to shift to a full smart grid system, it will have to be planned more comprehensively and rolled out more effectively than the smart meter programme. By mid-2018, it was reported that 11 million smart meter unit had been installed, but that the programme looked unlikely to meet its 2020 completion target and would only end up saving £11 per unit per annum, rather than the £26 at one time expected (Wilkinson 2018).

The smart meter story has yet to be resolved. It may turn out be a success. That is as yet unclear, but there are smart meter rolls out underway elsewhere in the EU and around the word, especially in Asia, although doubts have been aired about the need for some of them, at least at this stage (Holmstedt Pell 2016).

Unfortunately, while we wait to see what happens with smart meters, as we shall see in the next section, there is little uncertainty about some of system failures that the UK has already experienced in relation to more conventional energy-saving initiatives. Given that the focus in this book is on renewables, energy saving has not been central to our story, but it is clearly vital in terms of climate policy and emission reduction. Avoiding energy waste can cut costs and is often the easiest most cost-effective option for emission reductions. There are also direct links between energy saving and renewables: it makes it easier for renewables to supply the reduced demand. Moreover, if peak demand can be cut, then that can help with grid balancing, much like fully interactive smart meters, and possibly at less cost.

Sadly, despite the promise, the UK energy efficiency story so far has mostly not been a good one. The expected energy-saving benefits from smart meters, sometimes put at up to around 3%, may have so far been hard to get, but, while some of the other energy-saving schemes have been more hopeful of much larger saving, so far, with some exceptions, delivery has proved to be hard.

7.5 Energy Saving: the Green Deal

Energy saving is, or should be, very much part of the overall sustainable energy mix. Indeed, some say it should be the major part. Certainly since, given the availability of relatively cheap energy, energy efficiency had been mostly ignored, the potential savings are very large. However, once the easy savings (gathering the low hanging fruit) have been made, the cost of getting to higher levels rises and that may be only partly offset by economies of scale as the new energy-saving technologies are rolled out and improved. Moreover, even if, as is planned in Germany, demand can be cut by 50% by 2050 through energy efficiency measures, there would still be need for renewable energy to supply the other 50%.

Nevertheless, even given the problem of rebound (some energy savings will release money that will be spent on new energy-using services), saving energy remains a key economic, social and environmental issue, reducing emissions, saving money (and thus helping to relieve fuel poverty) and also offering important synergies with renewables: they can be mutually supportive (Elliott 2004; Herring and Sorrell 2009).

Successive UK governments have sought to address the issue of energy efficiency. All the studies have agreed that it was desirable, indeed vital, with major potentials for saving being identified. For example, a DECC report in 2012 'Capturing the full electricity efficiency potential of the UK' identified a potential for cutting energy demand by 40% by 2030, 155 TWh in all, of which current policy was estimated to capture about 54 TWh, around 35% of the total potential. DECC focused on three key sectors, residential, services and industry. In the residential sector, the potential saving from the top three energy-saving measures was put at about 58 TWh, via CFL lighting, appliances and better insulation, of which around 75% was expected to be captured by then planned policies (DECC 2012).

Some of that is being achieved, but, over the years, there has been more success in the industrial sector, in part because clear economic gains can be made in some industries by cutting energy use, especially in energy-intensive industries, where energy costs are the major operating cost. In most other sectors, however, progress on capturing major savings has often been poor, despite attempts to introduce support and incentive schemes. The mostly rather sad UK story on energy saving in the domestic sector was touched on briefly when we looked at the EST (see Box 5.1), an early (1990s) government initiative, which was cut back dramatically. Unfortunately, there have been many more examples of failed or minimised initiatives since then.

Perhaps the most notable example of a failure was the Green Deal scheme. The Green Deal had been prefigured by a plan announced in the Energy Bill of 2010 (DECC 2010b), under a Labour Government, with cross-party support. The Green Deal was a flagship manifesto commitment by the Conservative Party and was finally established by the Conservative-Liberal Democrat coalition government, and opened for business from January 2013. The initiative was led and strongly supported by Conservative Energy Minister Greg Barker.

The basic idea was to fund domestic energy efficiency improvements out of the savings from energy efficiency, with consumers repaying loans for upgrading their homes via their electricity bills. Lib Dem Energy Secretary Chris Huhne, who had also backed the Green Deal initiative, commented: 'The green deal means that home insulation is funded from the savings in home energy bills so people are better off, not worse off' (Huhne 2012).

Lib Dem leader, Nick Clegg, said 'we'll ensure customers are never charged more for the home improvements than we expect them to make back in cheaper bills. Plus the charge is attached to the property, rather than the person, so if you move, you stop paying. That is maximum affordability, with savings that should more than cover costs' (Clegg 2012).

The original plan envisaged 14 million homes benefitting by 2020. Homeowners would able to invest in energy-saving upgrades, using low-cost commercial loans offered by a range of private sector lenders for projects to be carried out by contractors under the new, separate Energy Companies Obligation (ECO), which replaced some earlier energy efficiency schemes targeted at energy suppliers (Ofgem 2018). ECO required energy suppliers to offer and support consumer energy-saving initiatives, with a levy on consumer bills being used to provide the money. The ECO scheme proved reasonably successful, although slow, with concerns being expressed that supply companies were not meeting the targets they had been given. Some utilities were fined as a result. It was thought that the Green Deal might stimulate more initiatives, with consumers incentivised to take action to save money.

Not everyone thought that was sensible or would work. For example, in an unusually frank open letter in 2011, the government's advisory Climate Change Committee said that, rather than in effect leaving action up to market pressures on consumers, it would be better to maintain pressure directly on the power companies and expand the ECO requirements: 'A higher level of ECO ambition would complement rather than crowd out Green Deal finance' (CCC 2011). There were worries about commitments to ECO, and other energy-saving measures, being reduced, with the Green Deal being used as a rationale.

Nevertheless, the Green Deal programme went ahead, although, unfortunately, consumer take-up of the loan offer was initially very limited. The painfully slow progress became something of a media joke. By

June 2013, it was reported that less than 10 consumers (perhaps 'as low as three') had signed up to the full thing, with repayment via electricity bills agreed, although, allegedly, maybe 200 more were 'in the pipeline'. The original plan evidently envisaged 10,000 in the first year (Bawden 2013).

Apart from some early IT glitches, one reason for the slow progress may have been the relatively high interest rate charged (6.96%) on the loans, along with a £20 upfront cover charge, and a £63 annual charge, all of which, it was estimated, could push the real interest rates on a financing package of £5000 up to 7.6% or more, depending on the duration of the loan. In addition, it seemed that around 20% of households would not qualify for a Green Deal package. Moreover, an initial poll found that only one in five people asked were aware of the Green Deal. That was addressed with a major advertising campaign, but take-up was still very sluggish, although evidently there was more success with Local Council-linked projects (DECC 2013b).

The government soldiered on, and some better progress was claimed to have been made after a revamp, with extra funding injected. In 2014, Liberal Democrat Energy Secretary Ed Davey said 'We're aiming to make 1 million homes warmer and more energy efficient by March 2015 – and we're on track with nearly 800,000 homes helped through ECO and the Green Deal by the end of August. More than 350,000 homes have now had a Green Deal Assessment and our recent research indicates that 70% of people go on to get work done', with DECC publishing some quite optimistic overviews (DECC 2014b, c).

The gap between those getting assessments done and those going on to agree to upgrade contracts began to reduce, especially following the introduction in July 2014 of a £120 million Green Deal Home Improvement Fund, with a 'fast cash back' grant and voucher system (DECC 2014d). However, as the numbers rose, the situation rapidly became almost farcical, with take-up of the grant scheme overwhelming the cash limits. The fund ran out. It had to be closed to new customers after six weeks (Gosden 2014b). Tragically, it then got worse. In December 2014, the Fund was re-opened, with an extra £24 million having been provided by the government for solid wall insulation projects, but it had to close again after just one day, following a rush of applications (Shankleman 2014).

It was clear that the system was not working. It needed more money to get it going fully. After a further £34 million government bailout loan in November 2014 failed to resolve its problems, and with still only 15,000 loan contracts agreed or in progress, in July 2015, DECC decided to no longer provide support, effectively ending the scheme (DECC 2015c).

In a retrospective review of the Green Deal initiative, the House of Commons Public Accounts Committee (PAC) said failures in the design and implementation had put public money at risk and must not be repeated. It said the take-up for the commercial loan scheme was 'woefully low' since it was not tested adequately. The forecast of demand for Green Deal loans was very optimistic, and 'gave a completely misleading picture of the scheme's prospects to Parliament and other stakeholders'. DECC had forecast that the Green Deal Finance Company would provide loans worth over £1.1 billion by end 2015, it was actually only £50 million. The Committee also raised concerns that, while taxpayers provided £25 million, more than a third of the initial investment in the Green Deal Finance Company, to cover set-up and operational costs, DECC had no formal role in approving company expenditure or ensuring it achieved value for money. The £25 million had now been written off.

Meg Hillier MP, Chair of the PAC, said: 'The Government rushed into the Green Deal without proper consideration of concerns about its weaknesses. Not enough work went into establishing the scheme's appeal to households, nor to its implementation, nor to examining the experience of governments setting up similar schemes overseas. This blinkered approach resulted in a truly dismal take-up for Green Deal loans and a cost to taxpayers of £17,000 for every loan arranged. Savings in CO_2 were minimal. Accountability to government of the Green Deal Loan Company – which spent public money on the expectation that it would need to support 3.5 million loans, compared to the 14,000 taken up – was institutionally weak. There is no doubt householders and taxpayers in general have been ill-served by these schemes and the Government must learn from its mistakes to ensure they are not repeated in this or indeed any other policy areas' (PAC 2016).

The Government's independent National Audit Office produced a similarly damning assessment. 'The Department did not set any expectations for the Green Deal. It did not state what proportion of measures' total cost should be paid for by the households that benefitted, either through Green

Deal finance or other means such as savings. Nor did it quantify the amount of CO2 the Green Deal should save in addition to suppliers' minimum obligations through ECO.' As a result of these and other weaknesses, 'Green Deal finance has saved negligible amounts of CO2', and overall, the Green Deal had failed to provide value for money (NAO 2016).

There are no doubt many ways in which this sad episode can be seen. While it did demonstrate that there was a strong appetite for energy saving, and some projects did go ahead, it has generally been seen as a disaster, including from a building industry perspective (Gardiner 2015). The scheme did seem to have been poorly thought out and run. Offering low-cost loans for cash-saving home improvements seemed likely to be popular, but the terms had to be right, and certainly, when they were adjusted, the applications accelerated, but to the point of overwhelming the extra funds that had been made available. In theory, once set up with initial government finance, it ought to have become a self-financing system, with income from consumer payments eventually offsetting the loans, but that would take time, with the Green Deal Finance Company, which handled the loans, carrying the debt meanwhile, aided in effect by a government float, some of which would eventually be paid back. However, even with large cash injections to try to get take-up to lift off, it never really got started. So some saw the whole thing as ill-conceived. Moreover, with the total government input put at £240 million, its failure fitted into the emerging political narrative about the alleged problematic nature and high cost of green initiatives.

The Green Deal's demise, in July 2015, followed the election, in May, of a Conservative Government, ruling alone without the Lib Dems. Clearly, it did not want to repeat mistakes like this. So perhaps unsurprisingly, the new government turned its attention to another green energy-related initiative, the Green Investment Bank (GIB). Although only partly related to energy efficiency projects (it also focused on renewable energy supply), it seemed clear that, as with the Green Deal, there was an emerging political pattern, a move away from the type of approach the Lib Dems (and before them, the Labour Government) had promoted. Ostensibly however, the initial problem was that GIB was having finance issues. The upshot was that it was sold off in 2017. See Box 7.8.

Box 7.8 The Green Investment Bank

The GIB had been set up by the coalition government in 2012, with £3.8 billion of public sector finance provided for investment in environmentally sound projects, including those which might not have sufficiently high short-term returns to attract private sector investment.

GIB was quite successful at financing selected new energy efficiency and renewable energy projects, in part since, as a public body, it did not have to use commercial rates of return, but could draw in private funding by reducing the risks they would face. So it backed a range of projects that might not otherwise have got funding, including, by 2013, 11 low-carbon wind and Energy from Waste schemes. In all, by 2015–2016, it had invested about £2.3 billion and attracted over £7 billion in private co-funding, rising to £3.4 billion of its own capital and £8.6 billion of private capital by 2017, this latter equating to around £2.50 for every £1 invested (GIB accounts 2012–16).

However, although GIB had broken even and began to make a working profit, its activities were limited by funding constraints. It could not, as originally set up, raise risk capital from the private sector so as to expand its investment portfolio: strictly, it was not actually a bank, but a government-funded investment agency. With the government being unwilling to provide more finance, what amounted to a partial privatisation was then proposed (BEIS 2015), with some seeing this as just a way for the cash-short government to raise money by selling off the profitable parts, others as way to allow GIB to expand.

A sell-off would certainly raise some cash, but allegedly the motive was that it would reduce risk to the exchequer and allow GIB to borrow money at commercial rates and allegedly get better 'value for money'. In a Parliamentary debate on GIB, the Business and Industrial Strategy Minister, George Freeman, said the change was needed since 'in Government ownership, the bank must compete for funding along with all other Government expenditure needs, in a necessarily tight public spending round. We do not want to constrain it because of that.' So the aim was to liberate it 'from the constraints of a tough public spending round' and allow it 'to go into the private markets and lead in the burgeoning green economy'.

By contrast, Green MP Carolyn Lucas asked if 'a profit-maximising Green Investment Bank will be unable to perform precisely that key role of reducing risk in important green sectors in order to crowd in private investment? There is a real risk that if the bank is put into the private sector it will crowd out other investors, rather than crowding them in.' Tory MP Graham Stuart asked 'what has changed so that such a bank can now be transferred to the private sector without ending up simply acting like and emulating all the other banks?' (Parliamentary debate 2015).

Nevertheless, after some further haggling, a full sell-off was proposed. That would effectively mean that any future profits would be lost to the taxpayer, and some felt that the green investment criteria it was meant to use would also be abandoned. That issue had been raised strongly by the Environmental Audit Committee (EAC 2015). In response, the government said it would create a special share in the company to protect its green mission, so it could veto inappropriate projects.

In 2017, GIB was finally bought by an Australian Banking Group, Macquarie Group Ltd, in a £2.3 billion sale. This led to some criticisms. The National Audit Office said the Government had sold it off too cheaply and that the buyer 'has no legal obligation to ensure GIB will keep focusing on its green objectives and be an 'enduring institution' for years to come' (NAO 2017). Clearly, it was not a very popular move (Vaughan 2018c).

Some may portray the demise of GIB, and the Green Deal, in party-political terms, as being just one result of a battle between the Lib Dems and the Conservatives, and a growing hostility to green initiatives. This may be over-simple. While the GIB was something of a Lib Dem flagship project, as we have seen, the Green Deal was initially backed by the Conservatives. However, views clearly changed. Conservative Prime Minister David Cameron, who initially seemed to back environmental policies, was later said to the have labelled the various schemes that had emerged under the coalition as costly 'green crap', which he wanted to get rid of (Sparrow 2013). But equally it might be seen in somewhat wider political terms; arguably, this is what happens when you try to use commercial incentives to deliver desired environmental outcomes. The results can be patchy or costly or both.

7.6 The Zero Carbon Home Programme

The privatisation of the GIB was clearly a shock for those who looked to it to provide accessible funding for green energy projects, including energy efficiency projects, while the Green Deal's demise was, arguably, disastrous for energy saving, adding to the list of energy-saving schemes that have been abandoned or watered down. Another tragic story can be told in relation to the 'Zero Carbon Homes' programme, which was also in effect wound up, although in this case, its demise was a matter 'death by a thousand cuts' over a longer period.

The zero carbon homes policy was first announced in 2006 by the then Labour chancellor Gordon Brown (Weaver 2006). The initial plan was for all new-build houses to be zero emission by 2016, with the emphasis heavily on energy saving by good building design and full insulation. To that end, a set of Codes for Sustainable Homes (CSH) were developed to guide new build, with increasingly high levels of energy performance being required (CLG 2010).

There had been many examples of 'low energy' houses and sustainable housing projects around the UK over the years, some of them experimenting with solar heating. The new city of Milton Keynes was one of the pioneers, initially benefitting from local development council funding and direction, with around 200 solar houses being built (NATTA 1987). But there were sometimes problems there and elsewhere, especially with the commercial projects that emerged when council house building programmes were halted under the Thatcher government in the 1980s. The performance and reliability problems (e.g. due to poor system installation and construction quality control) continued in the 1990s (Dougan 1998). So, arguably, there was a need for a more coherent regulatory framework.

The Zero Carbon Homes/CSH programme certainly represented a bold new approach. In the extreme, fully implemented, it could mean no external energy imports at all, with energy demand minimised and then met entirely by 'on house' renewables, presumably mostly solar, backed up by biomass and possibly micro-wind units where viable. It was specified that 'off-site renewable contributions can only be used where these are directly supplied to the dwellings by private wire arrangement', that is, directly from local sources. The aim was clearly to avoid imports of commercial energy via the grid, even if this was wholly or partly green energy.

However, after objections from house builders and developers, based in part on the allegedly high cost of meeting the new energy performance rules, and after a change of government (when Labour lost to the Con-Lib Dem coalition), the criteria and rules as outlined in the linked Codes for Sustainable Homes were progressively watered down. For example, on the demand side, it was decided to exclude power used for electronic devices and white goods (e.g. fridges) and that logically enough led on to exclude power used to power heat pumps. That could be a big concession.

Heating is the major part of domestic energy use. On the supply side, it was decided that imports from some external grid-linked renewables sources could after all be used, along with a range of other so-called allowable solutions, including carbon offsetting, that is, buying in carbon credits from projects like tree planting elsewhere. See Box 7.9.

Box 7.9 Undermining Zero Carbon Homes

The original aim of the Zero Carbon Homes policy had been for all the measures to be 'in' or 'on' house, following the so-called Merton rule, as developed by Merton Council, keen to ensure that building developers did not try to sidestep stringent energy-saving rules, for example, by relying on imported green power. Views differ on the wisdom of that, especially as applied to individual houses. It is clearly important for house builders to pay proper attention to energy saving/insulation in their house designs and to incorporate in/on house renewables where viable. PV can certainly help and so might solar heating and some other micro-generation devices, for example, heat pumps in off-gas grid areas. However, getting fully to zero carbon for heat *and* power that way would be hard and costly in many locations. Importing some top-up green power from more efficient large-scale wind farms in windy sites arguably would often make more economic and environmental sense.

The Conservative and Lib Dem coalition government's 2013 consultation on proposed changes to the scheme said 'the government recognises that it would not be cost-effective at this time, affordable or technically feasible to meet the zero carbon homes standard in all cases solely through measures on the dwelling itself, like fabric insulation, energy efficient services, and/or renewable energy generation measures (e.g. solar panels). Therefore the government proposes that house builders can achieve the zero carbon standard by mitigating the remaining emissions 'off-site', in effect a kind of carbon offsetting or abatement' (DCLG 2013).

It offered a three-tier approach, in-house efficiency, on/in/near house renewables, and bought in extra power/offsets. It said 'Requirements for fabric energy efficiency and carbon compliance are to be achieved by measures incorporated within or on the development site, including by direct connection to community energy schemes. They set a minimum threshold for onsite measures.' Note the inclusion of community schemes: it would no longer be necessary to do everything in/on one isolated house; 'connected measures (e.g. a heat network)' were also accepted for the core approach.

Arguably, linking into district heating networks might make more sense than domestic-scale micro-CHP in urban areas, so this seems sensible. But the key point was that the in/on/near house measures could be topped up with additional off-site 'allowable solutions', which could involve the house developer 'contracting with a third party Allowable Solutions provider for them to deliver carbon abatement measures sufficient to meet the house builders' obligations'. Moreover, it could also involve 'making a payment which is directed to a fund which then invests in projects which will deliver carbon abatement on their behalf. The payment would be based on a fixed price which would be subject to periodic review.' A big let-out clause!

It was proposed to leave it to the developer to choose the approach, although the consultation asked whether there should be a prescribed list to provide house builders with certainty and a clear indication of measures which are 'allowed'. It pointed out though that this might be inflexible and reduce 'the ability for new ideas to be brought forward and could stifle innovation'. There could be definitional and jurisdiction problems, especially for schemes like district heating, only part of which fed housing.

The government did look at the idea of having mandatory scheme under which 'Local authorities would collect Allowable Solutions payments from development in their area through a local levy and fund Allowable Solutions projects in their locality according to local priorities for carbon abatement', but it decided against recommending it, though local councils would be able to offer allowable solutions to developers.

As can be seen, the package represented some profound changes, some of them sensible, others arguably less so, in effect watering down the scheme. However, despite some objections, in its response to the consultation, the government accepted most of the proposals, although with verification and certification requirements and cost caps on the allowable solutions (DCLG 2014).

As an interesting coda to this, in 2014, the Government noted that it had 'not considered support for carbon capture and storage or nuclear power in the context of allowable solutions. However, the criteria based approach set out in the consultation for identifying appropriate carbon abatement measures would require that measures will need to bring forward additional, verifiable carbon savings at a cost effective price. Given carbon capture and storage, and nuclear power, would have high upfront costs, they would be unlikely to be cost-effective.' So here we have the government saying that nuclear and CCS are too expensive, at least in this context (Williams 2014).

As can be seen, if all these exceptions and allowable options were taken up, the 'Zero Carbon Home' might really be more like '50% carbon' or even much less, depending how you do the sums. The government's response to the consultation on changes to the scheme (see Box 7.9) talked of 'strengthened the energy performance requirements in the Building

Regulations for new homes', so that new homes would be 'required to be on average over 30% more energy efficient than before we came into office' (DCLG 2014).

Clearly, however, the scheme, as originally conceived, was on the way out, and in 2015, by which time the Conservatives were ruling alone, the zero carbon goal was in effect scraped, along with linked plans for new building regulations. The announcement on this was made as part of a government report, 'Fixing the foundations: Creating a more prosperous nation', which said the scrapping of the two regulations was designed to 'reduce net regulations on house builders'. Unsurprisingly, that was not well received by environmental groups (Oldfield 2015).

There was an attempt (via the House of Lords) to repeal the move, but that failed. A Department of Communities and Local Government spokesman reportedly said: 'We are already building some of the most energy efficient homes in the world. Our current standards are tough and already have the full support of the industry. Our aim is to speed up house building and not add extra costs and bureaucracy. We do not need extra legislation as existing legislation is already in place to allow energy performance standards to be set in Building Regulations' (Mark 2016).

Parts of the basic Code system have been retained, but the language used has begun to change to 'Low Carbon Houses', and the fully 'zero carbon' concept seems to have slid out of focus. Some might say that was sensible. It had been very ambitious, certainly when initially proposed, with 2016 arguably being very unrealistic for a 'zero' target date. However, the effective removal, or at least softening, of tough criteria for building emissions is not something to be celebrated.

There have been some new updated building regulations and other controls, and that is important for existing as well as new buildings, and, crucially, for rehabs, but the regulatory environment in the UK has arguably suffered much since energy market liberalisation and privatisation.

Energy saving in buildings is of course a quite lucrative area of business, with building fabric upgrades often being promoted as helpful as a way to reduce energy bills at low cost, although there is a technical debate over some aspects. For example, energy efficiency upgrades may not always be the best option for some building renovation projects. Indeed, some say that, if district heating mains are available, in some contexts, beyond a minimal weather proofing level, links to CHP-fed district heating networks can be a cheaper, more efficient and easier to install low-carbon option than substantial retrofitted insulation (Elliott 2015).

None of this means that energy efficiency in buildings or elsewhere is not still important or that governments are not seeking to improve it. For example, the ECO system is being revamped and, in 2017, the UK National Infrastructure Commission noted that energy efficiency measures could save nearly 130 TWh p.a., over a quarter of total UK electricity consumption. The UK government's 2017 Industrial Strategy included a £162 million innovation programme, with some of that going to support investment in industrial energy efficiency (BEIS 2017a). Hopefully that will all have an impact. Meanwhile, the European Commission has produced many useful studies on what can be done, along with a policy commitment to increase overall energy efficiency in the EU, with energy consumption to be cut by 30% by 2030. Whether the UK will take notice of that after BREXIT remains to be seen.

7.7 Green Heat

The problems faced by attempts to improve energy efficiency in the UK have been paralleled by the problems it has faced in trying to expand the use of green heat. That too is clearly needed to get emissions down, and can also help with grid balancing. Moreover, although it is about *supply*, providing heat for buildings cannot sensibly be considered in isolation from measures, such as insulation, to reduce heat loss and energy *demand*. The Renewable Heat Incentive, set up in stages from 2011 (see Box 7.1), required that this was taken into account before grants for green heat supply were agreed. However, as with the Green Deal, take-up under the RHI was slow.

The National Audit Office noted that just under 80,000 installations had been put in place by the end of 2017, and it was estimated that, in all, only around 111,000 new installations might be completed by March

2021, only 22% of the expected 513,000 by 2020 (NAO 2018). Ofgem also estimated that 4.4% of non-domestic schemes were non-compliant during 2016–2017, with about £3 million in overpayments to RHI participants resulting. Fraud problems had emerged with the RHI in Northern Ireland (which had its own, evidently poorly drafted, version) leading to some major political problems (BBC 2016).

Problems like this aside, the overall cost of the scheme, which falls on taxpayers, was quite high. The RHI provides an incentive for businesses or home owners via a tariff over a 20-year period for each unit of heat produced from renewable sources, with total payments between November 2011 and August 2017 of £1.4 billion. It was expected that outgoings to new participants could reach £23 billion by 2040–2041, if the scheme continued. However, at present, the RHI is due to end in 2021, and there is no clarity on what happens, as regards new projects, after that.

The scheme's prospects, unless modified, do not look too promising. In 2018, the Public Accounts Committee produced a very critical report claiming that 'the Department's forecasts of take-up were wildly over-optimistic'. As a result, 'the Department has had to cut back its expectations of how much renewable heat will be produced by the scheme by almost two-thirds, and of the reductions in carbon emissions due to the scheme by almost half. This means other policies are having to work harder to enable the government to meet its legal obligations' (PAC 2018).

Issues of project and payment monitoring apart, some of the criticisms concerned the choice of technology. The emphasis was on biomass and heat pumps, in line with the government's decarbonisation policy. These might make sense for off-gas grid areas (about 15% of UK homes), which were the main focus for the domestic RHI. But solar heating could play a role in most places. Although solar heating was initially backed with quite high RHI tariff support (19.74 p/kWh), there were proposals to drop it entirely from the RHI, on the argument that it was too expensive (BEIS 2016b). Understandably, this did not go down well with the STA (STA 2016). BEIS eventually relented, arguing that 'while the tariff in support of solar thermal is still high compared to some other tariffs in the scheme, the possibility for continued support to deliver cost reductions suggests the long-term value for money of this support will be better than previously thought. In addition, the role of continued support in maintaining the UK supply chain, particularly with regard to UK based manufacturing of solar panels, improves the value of continued support' (BEIS 2016c).

It may be true that solar heating is not currently competitive for individual home units in the UK, if gas is available, but larger shared solar heat collector systems, with large shared heat store backup, might be more attractive in many locations, possibly linked into local heat networks. The UK STA noted that 'One of the largest solar thermal schemes in the world, in Silkeborg, Denmark, provides a fifth of the heat for the district heating for a town of 43,000' (Greene 2017). Something similar has been tested in the UK, with E.ON's solar-thermal district heating and heat store/heat pump 'Project Sunshine' in Devon (Appleyard 2016).

There are similar issues of scale in relation to biomass. Biomass was strongly supported in the RHI initially, at all scales, including large industrial units, but in 2015, there were proposals for cuts of up to 45% in support for some projects, including domestic-scale boilers. That was mostly fought off, but there were reforms to the RHI, with some biomass projects being constrained. Some of this was due to arguably reasonable environmental concerns. For example, there can be problems with toxic emissions from poorly installed and managed domestic stoves/boilers, and in relation to the non-domestic biomass RHI, BEIS was worried about crop-based feedstocks being used for biogas production in Anaerobic Digestion (AD) plants: 'The primary purpose of agricultural land should be for growing food. However, growing deployment of AD on farms has caused a significant increase in the use of crops for AD. In 2014, maize grown for AD made up 19% of the total maize area in England and 0.7% of England's total arable are a significant portion of planned plants whose operations are geared towards high crop use. In particular, the Government has specific concerns about the impact of late harvested crops such as maize on soil and water quality' (BEIS 2016b).

Although it recognised that there might be some cases where surplus crops could be used, BIES was more interested in using farm and food bio-wastes as a feedstock. Its RHI reforms included the requirement for new AD plants 'to produce at least half their biogas and biomethane from waste-based feedstocks to receive support for all their production. This will help divert wastes from landfill and make use of available resources' (BEIS 2016c).

More generally, 'value for money' and longer-term development criteria were imposed on the reformed RHI. That led to some debate. For example, domestic heat pumps had faced some installation and operational problems, although BEIS continued to support them in the RHI reforms. However, some felt that, while domestic-scale systems like this, and also individual domestic biomass-fed boilers, might have a role in some locations, larger community-scaled systems, possibly CHP units, with large heat stores, linked to district heating networks, were likely to be more efficient and better managed. Green gas (biogas or hydrogen) was a parallel option, of probably wider relevance, since not everywhere could be serviced by heat mains, with the key attraction of biogas (and possibly hydrogen) being that it could be delivered using the existing gas mains and did not require significant changes in end-use devices: there was no need to install expensive heat pumps.

This line of argument, pointing to the need for heat networks, fed by waste industrial heat, or by biomass or solar derived heat, and/or for the use green gas delivered by pipe, rather than individual heating units like heat pumps, essentially amounts to a challenge to the 'decarbonisation by electrification' model initially promoted by the UK government (see Sect. 7.1). Support for that new view, or versions of it, emerged from a range of sources across the political spectrum, with the generally right-wing Policy Exchange unhappy with what it saw as a wasteful electrification approach (Gosden 2016) and, like some left wingers, being keen on green gas (Labour 2017). Interestingly, the government's advisory National Infrastructure Commission also backed green gas/hydrogen quite strongly (NIC 2018), with the study done for it by Aurora suggesting that 'hydrogen based heating puts less strain on the electricity system' (Aurora 2018).

Pushing the green gas case even harder, David Smith, Energy Networks Associations CEO, said 'the NIC figures suggest that a Whole-System Approach to decarbonisation, which involves planning how to use our energy network infrastructure in a more integrated way by using low carbon gasses like hydrogen, could save £55 billion to 2050 compared to an electricity only solution' (Smith 2018). In parallel, on the heat network side, the Energy Technologies Institute said nearly *half* of existing UK heat demand could be connected economically to green heat supply networks, up from 2% now, saving £30 billion (ETI 2017). City-wide CHP/district heating had of course long been popular on the left. Labour had backed plans for city-wide CHP/DH in Newcastle, Sheffield and elsewhere in the 1970s, but the programme was not taken up. However, with heat networks now back on the agenda, and green gas now being pushed by the H21 hydrogen project in Leeds, one way or another, it did look like the 'electrification' route was dead, or less central.

If that view wins out, as seem possible, then the RHI may have to be reconfigured further and linked with an enlarged green heat network/ green gas programme. As noted earlier, the government has provided £320 million to stimulate heat network development. That is small, but it is expected to draw in up to £2 billion of additional capital investment and lead to the construction of hundreds of heat networks in England and Wales. In parallel, Scotland has set an ambitious target to connect 40,000 homes to heat networks by 2020, representing 1.5 TWh of Scotland's heat demand. It is part of Scotland's ambitious plan to get 50% of all its energy from renewables by 2030 (Scottish Government 2017).

Clearly however, a lot more effort will be needed to meet high heat targets, with biogas production from wastes also being stepped up, rather than being partly sidelined. In its initial report on energy networks in 2014, DECC had said 'innovative alternatives to natural gas, such as biomethane, low carbon hydrogen and gas from anaerobic digestion plants are potential ways of maintaining the existing gas grid while reducing our reliance on fossil fuel gas' (DECC 2014e).

As we have seen, biogas does have some RHI support, with the focus on using bio-wastes. In the RHI reforms, DECC argued that 'where food waste cannot be prevented, AD is the best available treatment option. AD can reduce emissions on farms where it is used to treat manures and slurries, as well improving the management of on- farm nutrients and providing a route for farm diversification' (BEIS 2016c). However, with the block on using crops, the volumes might be lower. It may be recalled (from Chap. 6) that, in parallel, biogas AD was also being constrained by the FiT cuts for small power generation projects, some of these being micro-CHP based, so that the heat supply side might also be impacted.

The other green gas option, Power to Gas 'green hydrogen' production, is still an outrider in the UK, but arguably needs to be fully explored.

Overall, if all this was followed up, and some of the constraints were removed, it does seem like there could be a viable green heat supply package, with green gas playing a major role (Abbess 2015). In addition, there is the direct heat network option, with CHP being a low-carbon heat source and zero carbon if using biomass. As noted earlier, solar can also play a role, and in some locations, geothermal heat could also be fed in. Some new geothermal heat projects have been proposed in Manchester and Newcastle and also in Cornwall. A project was also mooted in Stoke on Trent.

Not everyone agrees about all aspects on this new agenda. See, for example, the cross-party report 'Policy for Heat: Transforming the System' (Policy Connect 2015) and the review of practitioner opinions carried out by the trade magazine *The Energyst*, with one contributor saying 'heat networks would require subsidy of 75p/kW hour to create the seven-year payback periods sought by investors', which 'would make even Hinkley C look cheap' (Energyst 2016). Some are also less enthused about biomass use than others, given the environmental issues. As noted above, BEIS had concerns about using crops for AD biomass production, and for biomass generally, getting the sourcing right is obviously crucial, to minimise carbon sink loss. Even so, clearly there will be land-use constraints. As I have argued in an earlier study, there may not be enough biomass available for all end uses, especially initially (Elliott 2007), so it is important to decide on the best end use. For example, should biogas be used primarily for heating, perhaps injected into the gas grid? Or should it be used for farm tractors, or trucks and vans? Or should it, and other types of biomass, be burnt to make electricity? The same issues may emerge for hydrogen: it might not be possible to produce enough at low cost for all end uses—power, heat and transport, as well as balancing.

There are certainly debates on the right mix of options and end uses, fed by some helpful academic and research institute reports on green heating (Douglas 2015; MacLean et al. 2016) and more recently by National Grids Future Energy Scenario studies, mentioned at the end of Chap. 6. Their most recent version offers a range of high renewables scenarios, with some dominated by hydrogen, others by electric heat pumps, with heat networks playing a background role (FES 2018). So choices remain. For example, for heating, it may well be that a hybrid approach, with domestic heat pumps being used in some locations, green gas and heat networks in others, may turn out to be the cost-effective optimal, with, in one study, the 'hydrogen grid alone' approach being seen as the most expensive (ICL 2018). Although another study seems to have come to different conclusions (the hydrogen route might be the lowest cost) and it also saw a role for heat networks (Element Energy and E4tech 2018). Certainly, along with hydrogen/green gas, the heat network idea seems worth serious emphasis. An early DECC estimate, when launching its proposal for nine tests projects, was that 14% of UK heat demand could be cost-effectively met by heat networks by 2030 and around 43% by 2050 (DECC 2015d).

Looking to the future, DECC said the new system 'will involve a great expansion in heat networks to meet our heating needs, allowing for the exploitation of secondary heat sources such as recovered heat from industry, power generation, infra-structure and waste processing, as well as direct renewable heat like deep geothermal, and water-sourced heat pumps. These technologies and methods all exist today but so far are limited to small pilots or vanguard projects. They will need to become mainstream. In addition, innovative alternatives to natural gas, such as biomethane, low carbon hydrogen and gas from anaerobic digestion plants are potential ways of maintaining the existing gas grid while reducing our reliance on fossil fuel gas' (DECC 2015e).

Although, as we have seen, BEIS, DECC's replacement, has shown continuing interest in these options, there is a way to go on green heat, and some choices to be made, but at least a start seems to have been made, with some positive outline visions of what might be aimed for and the beginnings of a practical programme.

7.8 What Next for the Energy Transition?

Dealing with heating is important, and some of the measures looked at above can help with system balancing, but that will only be part of the solution to the wider system balancing problem. The grid balancing issue is quite complex, and it has taken time for the government to address it.

As this chapter should have indicated, not all its solutions have been seen as wise or likely to be effective without some changes, as with the capacity market, for example. Some of its other green energy initiatives have been non-starters, or have been abandoned. However, there are signs of progress and indeed enthusiasm, as in DECC's almost evangelical support of smart grids and 'prosumer' power. See Box 7.10.

Box 7.10 A Smart Energy Vision

Offering support for a full smart power grid system and backing the rise of self-generating prosumers, in 2015, DECC suggested that 'consumers will become active players in the energy system. In addition to generating and producing their own energy, through dynamic "time-of-use" tariffs, consumers will be able to benefit from cheaper energy by using it away from peak times. The smart home of the future will allow consumers to use smart meters and home generation to automatically set smart appliances to use energy at the cheapest time of the day – ensuring convenience and reduced energy bills.'

It also looked to a future economic transformation: 'The market will no longer be dominated by a small number of large energy companies and traditional business models. With an increasingly diverse energy mix, there will be more opportunities for smaller, non-traditional players to participate in generating, storing, transporting and selling energy, maximizing competition and bringing much needed diversity to the system, helping to reduce cost overall. Increasingly, communities will come together to manage their energy needs locally, building their own generation, managing their energy use to maximise the use of locally produced electricity and using revenues to invest in energy efficiency and local amenities. This will enable communities and other smaller players in the market to establish local markets where they buy and sell their own energy and become increasingly self-sufficient.'

It concluded: 'This revolution in the energy system will need a transformation in our energy networks. Smart grids will be at the heart of this, allowing us to manage this changing energy system more efficiently and at lower cost, providing a more resilient network that enables a much more diverse range of players into the market. The local network operators will take a more active role in managing supply and demand at the local level, as National Grid currently does at the national scale. The smart grid transformation is already underway, but in the future we will need a smart energy system that is able to integrate the use of electricity, heat and gas to maximise efficiency, reliability and reduce costs' (DECC 2015e).

DECC may have gone, but it is hard not to be inspired by this vision, which almost stands comparison with some (but not all) early AT rhetoric. Similarly, ambitious visions have emerged from, amongst others, the UK Institute for Public Policy Research, looking at what can be done to shift to decentralised power. It focuses on 'prosumer' and community initiatives, aided by digital system integration and a flexible energy system which 'empowers citizens and communities to be more self-sufficient while being part of a connected, inter-dependent system that offers security of supply and resilience in the face of changing demand and climate' (Lloyd 2018).

In addition to the likely social and environmental benefits, the potential economic gains claimed from smart flexible systems certainly look enticing. For example, a study by Imperial College London and OVO Energy, a Bristol-based energy supply company, 'Blueprint for a Post-Carbon Society', has calculated that, in a very high UK renewables scenario (93% of electricity by 2040), the use of residential flexible technologies, such as smart EV charging, smart electric heating and in-home battery storage, could save the UK energy system £6.9 billion overall. It says 'Flexible storage, located near consumption and found in EVs, smart electric heating and home energy storage devices offer a perfect solution to ease grid capacity issues and will limit the need for expensive grid upgrades and reinforcements. The energy storage found in these behind-the-meter (BTM) devices can act like an energy reservoir, soaking up cheaper renewable power that can then be used when required or released back into the grid at times of peak demand' (OVO/ICL 2018).

The report suggests that the results show that smart electric heat can provide enough flexibility to enable green generation from wind and solar alone, displacing the need for nuclear and CCS. The savings were calculated as £1.1 billion from smart EV charging, £3.5 billion from V2G EV charging, £3.9 billion from smart heating systems and £2.9 billion from in-home batteries. That would be equivalent to a £256 saving on the average household energy bill each year. The scenario relies on the uptake of 25 million EVs and 21 million electric heating units by 2040, which the report says is 'ambitious but achievable' (Edie 2018a).

Even allowing for hyperbole, there is much that is attractive in the visions and scenarios outlined above. However, they are still some way from reality. For example, the problems with the smart meter roll-out have continued (Pratt 2018b), and that experience suggests there could be even more implementation issues with the full development of a smart energy system, which would cut across many vested interests, sectors and technologies, including EVs and V2G. So, although well worth exploring seriously, it may be wise to be a little wary of too much overhyping of smart grids and complex interactive systems. A rapid move to flexible systems seems to be the right direction of travel, as *UK Power Networks* have argued (Edie 2018b), but there is still some way to go, and it may take time to develop and deploy all the new systems fully.

It may be wise to be even more wary of those overhyping the role of nuclear power, whether for electricity supply and grid balancing, or, as has also been suggested of late, for direct heat supply, or the production hydrogen for heating or vehicle use. Indeed, some portray these possible new applications as vital to make nuclear more economic, given that it is having trouble competing in the electricity market (Elliott 2017b).

Nuclear certainly has problems. While some (including the UK government) remain enthused about the 'new nuclear' programme, as costs have risen, there has been a growing level of criticism. The *Economist* was one of the first 'establishment' orientated outlets to break ranks, commenting, in 2016, in relation to the proposed Hinkley plant, that 'Britain should cancel its nuclear white elephant and spend the billions on making renewables work' (Economist 2016) and subsequently continuing to report on the global financial problems faced by nuclear (Economist 2017). The economic problems have certainly become severe and have added weight to the arguments used by those who oppose nuclear power, including those who see it as undermining the development of renewables, which they view as a better option (No2 Nuclear 2018). See Box 7.11.

Box 7.11 Nuclear Versus Renewables Funding

Those involved with the incumbent energy technologies like nuclear usually have been less than enthusiastic about renewables, with the nuclear lobby often in the early days being quite disparaging of them. That stance gradually changed as renewables began to achieve success, with the new position typically being that 'we need both': they can work together. In that context, it was a little hard to explain why nuclear was getting the lion's share of the funding, in an over 10:1 ratio up until to the 1990s (it ranged from 12:1 to 19:1 between 1979 and 1985: see Table 3.1). In one attempt to provide an explanation, in 1988, Energy Minister Michel Spicer argued that 'there is an inherent difference in scale, and hence in costs, between the R & D needed to achieve results in the different sectors. Nuclear R & D is at the development stage, which is also inherently more expensive than the research stage, at which much of the non-nuclear work is being done at present. The Government look at each research area and programme on its own merits, in the light of advice from ACORD, and allocate resources accordingly, rather than try to seek a spurious balance between them' (Spicer 1988).

In the event, following privatisation, nuclear R&D, indeed all energy R&D, fell dramatically, with one result being that, from around 2001, nuclear and renewables R&D funding reached something like parity, with most of the government's nuclear R&D allocation going to nuclear fusion (DTI 2001). That is not to say nuclear power has not continued to benefit in other ways, with a range of support initiatives, as well as the CfD for Hinkley, but so did renewables, even if, as we have seen, debates continued over the comparative levels of support. The simple pro-renewables argument is that, with the overall energy budget being tight, every pound spent on nuclear is a pound not spent on renewables, with the latter giving better value for money, and also avoiding major future radioactive waste management costs, proliferation risks and security issues. The pro-nuclear argument seems to rely mostly on the belief that new technology will be cheaper, a claim that has been made throughout most its long development period, but a promise that has yet to be fulfilled.

While most still see nuclear and renewables as clear rivals, the wider debate over nuclear continues. A subset of it concerns its role in grid balancing and also potential new roles as a heat and possibly hydrogen supplier. The heat supply option would require the development of smaller nuclear plant that could be installed in or near cities. Some in the nuclear lobby are certainly pressing for SMRs with that in mind. However, there would be major problems, not least in relation to safety and security.

Opinion surveys have already indicated that few people would be willing to live near such a plant (Vaughan 2017).

Hydrogen production by electrolysis of water is a possible option, as with P2G, providing a storable fuel, but using surplus renewable power for this purpose is likely to be cheaper. Hydrogen production by direct dissociation of water is more speculative, possibly requiring the development of new high temperature reactor types.

For its part, the government has shown only limited interest in supporting new nuclear technology, with, in 2017, just £4 million being allocated for feasibility studies, followed, possibly, depending on the results, by £40 million for R&D on 'advanced modular reactors' (WNA 2017). Moreover, the level and range of support subsequently proposed for SMRs, in the BEIS SMR competition programme, seemed to be somewhat more constrained (Dawson 2018). All in all, new UK nuclear technology seems a very long shot. Some however insist that conventional nuclear plants could play a role in grid balancing, and more generally in energy supply, although many see large inflexible nuclear plants as being of little use for balancing variable renewables. See Box 7.12.

Box 7.12 Nuclear Balancing

As renewables have begun to dominate, the nuclear lobby has sometimes fallen back on the argument that nuclear plants are needed to back up variable renewables. It is true that nuclear plants can vary their output to some extent, for example, to follow the slow daily cycles in demand, and they have done so in France and Germany. However, this undermines their economics. Nuclear plants are usually run flat out, 24/7, to recoup their large capital costs, and there are technical and safety issues which make it hard for them to vary their outputs rapidly and regularly, as they would have to do to balance variable renewables. Some may try, although there are clearly limits (Morris 2018).

As things stand, it is not planned that any of the new reactors proposed in the UK will load follow. That may change with new reactor designs, but, as we have seen, there are easier ways to balance variable renewables than by building expensive new nuclear plants. Indeed, having large basically inflexible nuclear plants on the grid gets in the way of the flexible balancing system that is needed. In particular, it can increase the need for overall output curtailment when energy demand is low. For example, UK power demand can fall to around 20 GW or less at night in summer. With demand for electricity actually reducing in the UK, National Grid has suggested that peak demand on summer nights might only be 17 GW in future. If UK nuclear capacity is expanded as planned, it could meet most of that. So whatever output was available from the large amount of renewables capacity that should be in place by then would at times be surplus to requirements. Or would the nuclear plants be shut down briefly to make way for the renewable power, which is likely to be cheaper? Neither option makes much economic or technical sense. And the conflict would deepen as more renewables came on line.

What seems to be indicated is that, unless there is a very large (and expensive) storage capacity, nuclear and renewables do not fit well together on the same grid system. Export overseas is also an option for some of the surplus, but night time demand is likely to be low everywhere. In which case, rather than building large new inflexible nuclear plants and curtailing renewable output regularly, arguably, the UK should be focusing instead on renewables, combined with flexible balancing systems. That does not mean there is no role for some nuclear. For example, some say SMRs may be more flexible. However, they seem some way off, with a range of economic, safety and security issue to address (Elliott 2017b; POST 2018).

At present, the UK is still going for a non-fossil mix including a substantial and expanding nuclear contribution. That is rare in the EU. Nevertheless, the nuclear lobby can point to China and India as examples of other countries that are still pressing ahead with nuclear, alongside renewables. However, in both those cases, wind-power output has already overtaken the output from nuclear, and, with PV also now booming, renewables look set to continue to dominate. Certainly, with hydro included, renewables supply around seven times more electricity than nuclear in China (WNISR 2018).

Nuclear plants do not directly generate carbon dioxide, but producing the fuel for them does (it is an energy intense process), and, although new extraction and fabrication technologies may be less energy intense, that requirement will increase as reserves of high-grade uranium are depleted. More energy will be needed to produce the fuel, possibly expanding ultimately to the so-called point of futility, when the energy requirement for fuel production begins to match the energy the nuclear plant produces. Longer term, given this possibility, and, more immediately, given the high cost of nuclear and its many other issues, it could be that, in countries like China, the commitment to nuclear will reduce, as, interestingly, it has, post-Fukushima, in Taiwan and also South Korea. Given China's huge size and weak grids, it is certainly hard to see how the expansion of large-scale centralised nuclear could avoid conflicts with decentralised renewables, if the latter continue to expand rapidly.

The UK situation is similar, although it has the advantage of a well-developed grid system, so balancing variable renewables and avoiding curtailment should be easier, if it avoids large-scale nuclear expansion. However, that does mean that there will have to be a change in viewpoint. The 'baseload' concept, with a few large power plants meeting most demand most of the time, has to be replaced by the now arguably more relevant concept of having a flexible supply and demand system, with multiple heat and power inputs at a range of scales and locations (Holliday 2015; Brattle 2017). Signs are that this idea is beginning to be taken onboard, but as the account above of past experiences has indicated, it may take time to bring about the necessary changes.

Major change will certainly be necessary, not just in terms of grid balancing, but for the whole energy transition programme, if the UK is to meet its climate targets. In 2016, the Energy and Climate Change Select Committee said the UK would fail to reach its legally binding target of providing 15% of its energy from renewables by 2020, due to 'under-performance' in heat and transport. It noted that the UK was not even halfway towards achieving the target of getting 12% of energy for heat generation from renewables, while the proportion of renewable energy used in transport had in fact fallen, from 4.9% to 4.2% in the past year (Energy and Climate Committee 2016).

The situation may have improved a little since then, although, in its 2018 report, the Committee on Climate Change was not much more optimistic (CCC 2018). While renewable electricity generation had been doing well, led by offshore wind, there were still major problems in terms of heat and especially transport. We have looked at some of the new moves on green heat above, and as was noted, there are many potential low-carbon technical fixes for transport, including electrification, the use of biogas, hydrogen and syngas, as well as power balancing options to avoid grid stress from EVs (FES 2018).

However, a key problem is that demand for mobility keeps rising. We are chasing a moving target. There is a risk that rising transport energy demand, especially for powering private vehicles, will drive the development of the energy supply system, for example, using more and more hard won and valuable green power for EVs and also perhaps diverting biofuels, biogas and hydrogen away from arguably more important end uses, such as heating and grid balancing.

In 2018, the government announced a new expanded 'Road to Zero' programme to support low emission vehicles development and uptake (HMG 2018), and EV uptake is certainly expanding, but arguably there is a need for a radical rethink of mobility issues, something well outside of the remit of this book, as is the air transport issue. That is also true of the wider issue of taming energy demand. We have looked at energy efficiency measures, and clearly the potential there is very large. One recent study suggested that 'one quarter of the energy currently used in UK households could be cost effectively saved by 2035; and this could increase to one half if allowance is made for falling technology costs and the wider benefits of energy efficiency improvements' (Rosenow et al. 2018).

In theory, even larger savings might be obtained from more radical changes in energy use and lifestyle and by the wide acceptance of the idea of sustainable consumption. I will look some more at that large issue in the next, final, chapter. It takes us back to the start, to some to the ideals promoted by the 1970s AT and ecology movements.

However, by way of an interim conclusion to the technological parts of this book and to this 'what next' section, in a more focused way, if I had to identify what I think could be the key technical issues/questions for the next phase, my (annotated) list would go something like this, mostly concerning likely 'game changing' breakthroughs:

1. Will solar PVT dominate? Hybrid solar-thermal systems are attractive, since PV cell performance falls with rising temperature. So it is good if heat can be collected by the same device for parallel use. But will PVT become used widely at all scales?
2. Will floating offshore wind triumph? Floating system can be located far out to sea, regardless of water depth, and avoid the need for

expensive sea-bed foundations. So they can run where the wind regime can be very good. But since the power has to be brought in over a longer distance, will this approach win out?

3. Will flying wind work? Aerial devices, with rotors mounted on flying wings or balloons or kites, or kites pulling winch cables connected to ground-based generators, can tap into the much higher wind speed at altitude. Will getting access to the bigger resource up there compensate for the problems, for example, the risk of crashes, collisions?
4. Will tidal lagoons win out? Tidal barrages are environmental invasive, but lagoons need not be, since they do not block the entire estuary. But will they be economic? Or will free-standing tidal current turbines be better?
5. Will wave energy ever be viable? A new obituary has been produced in the UK: is it premature? Is it just too hard to capture energy from the chaotic interface between air and water, compared to the smooth undersea tidal flows?
6. Will geothermal energy ever be viable on a large scale. There are good 'deep heat' resources in some UK locations, but getting reliable access to them has proved hard.
7. Will Power to Gas become a major grid balancing option? Producing hydrogen using surplus wind/PV derived electricity, and storing it ready to generate power when wind or PV is low does seem a clever solution to intermittency. But will it be economic?
8. Will BECCS, Biomass with Carbon Capture, ever become viable? It could be a negative carbon option. However, it would need large biomass plantation areas and also large-scale CCS, an option which so far has not been developed.
9. Will heat networks spread? CHP has many attractions and can use a range of sources, but district heating is not possible in all areas and its reliability can be problematic.
10. Will 'prosumer' self-generation by domestic consumers spread, as it has in Germany, mainly using PV solar and sometimes now also batteries? New smart grid tariffs might help, but there can be problems integrating this power into the overall system.

11. Will Block chain systems help us to manage smart grids and peer-to-peer trading of renewable power without incurring huge energy debts? Some bitcoin ledger-type systems are claimed to be *very* heavy power users.
12. Will long distance HVDC supergrids spread to help balance variable renewables and variable demand over wider areas? Perhaps even importing power from North Africa?
13. Will SMRs ever become economically available and socially acceptable, so as to get nuclear power back in the race? They might try to compete with biomass-fired CHP for heat supply, but would anyone want one in or near a city?
14. Will nuclear fusion become viable? It still seems a long shot with many unknowns. Why not use the free fusion reactor we already have—the sun? It already supplies 25% of our power, and as conversion technologies expand, it can supply much more.
15. Will we ever learn to use energy, however produced, efficiently? The economic and climate benefits of energy saving are clear, but the emphasis has so far been mainly on increasing energy supply. There are many technical options, but we also need to think in terms of sustainable consumption.

References

Abbess, J. (2015). *Renewable Gas*. Basingstoke: Palgrave.

Ambrose, J. (2017, December 4). Energy Networks to Unveil Plan for £17bn Smart-Grid Boom. *Daily Telegraph*. http://www.telegraph.co.uk/business/2017/12/04/energy-networks-unveil-plan-17bn-smart-grid-boom/

Appleyard, D. (2016, August 31). Hybrid Solar Thermal-Heat Pump on Trial. *Renewable Energy Focus*. http://www.renewableenergyfocus.com/view/44671/hybrid-solar-thermal-heat-pump-on-trial/

Auffhammer, M., & Wolfram, C. (2018, January 8). *Bitcoins Should Be Called BTUcoins, and That's a Problem*. Energy Institute at Haas Blog. https://energyathaas.wordpress.com/2018/01/08/bitcoins-should-be-called-btucoins-and-thats-a-problem/

Aurora. (2018). *Power Sector Modelling: System Cost Impact of Renewables*. Report for the National Infrastructure Commission, Aurora Energy Research.

https://www.nic.org.uk/wp-content/uploads/Power-sector-modelling-final-report-1-Aurora-Energy-Research.pdf

Bawden, T. (2013, June 23). Government's Green Deal Branded a Failure as Fewer Than Ten UK Homes Take Out Loans Offered. *The Independent.* https://www.independent.co.uk/environment/green-living/government-s-green-deal-branded-a-failure-as-fewer-than-ten-uk-homes-take-out-loans-offered-8669959.html

BBC. (2016, December 23). RHI Scandal: RHI 'Cash for Ash' Scandal to Cost NI Taxpayers £490m. *BBC News.* https://www.bbc.co.uk/news/uk-northern-ireland-38414486

Beez, W. (2017, October 13). The Future of Power-to-Gas Couldn't Be Brighter. *Renewable Energy World.* http://www.renewableenergyworld.com/articles/2018/02/the-future-of-power-to-gas-couldn-t-be-brighter.html

BEIS. (2015). *The Future of the UK Green Investment Bank plc.* London: Department for Business, Energy and Industrial Strategy. https://www.gov.uk/government/organisations/uk-green-investment-bank

BEIS. (2016a). *New 'Central Heating for Cities' to Help Reduce Energy Bills.* London: Department for Business, Energy and Industrial Strategy. https://www.gov.uk/government/news/new-central-heating-for-cities-to-help-reduce-energy-bills

BEIS. (2016b). *The Renewable Heat Incentive: A Reformed and Refocused Scheme.* London: Consultation Document from Department for Business, Energy and Industrial Strategy. http://www.gov.uk/government/uploads/system/uploads/attachment_data/file/505127/The_Renewable_Heat_Incentive_-_A_reformed_and_refocussed_scheme.pdf

BEIS. (2016c). *The Renewable Heat Incentive: A Reformed Scheme.* London: Department for Business, Energy and Industrial Strategy. http://www.gov.uk/government/uploads/system/uploads/attachment_data/file/577024/RHI_Reform_Government_response_FINAL.pdf

BEIS. (2017a). *Industrial Strategy: Building a Britain fit for the future* (White Paper). London: Department for Business, Energy and Industrial Strategy. http://www.gov.uk/government/uploads/system/uploads/attachment_data/file/662541/industrial-strategy-white-paper-print-version.pdf

BEIS. (2017b). *Upgrading Our Energy System: Smart Systems and Flexibility Plan.* London: Department for Business, Energy and Industrial Strategy. http://www.gov.uk/government/publications/upgrading-our-energy-system-smart-systems-and-flexibility-plan

BEIS. (2017c). *Heat Network Investment Project*. London: Brochure, Department for Business, Energy and Industrial Strategy. http://www.gov.uk/government/uploads/system/uploads/attachment_data/file/691643/Heat_Network_Case_Study_Brochure.pdf

BEIS. (2017d). *Smart Meter Customer Experience Study*. London: Department for Business, Energy and Industrial Strategy. http://www.gov.uk/government/uploads/system/uploads/attachment_data/file/641076/Post-install_key_findings_report_FINAL_24082017_PUBLICATION.pdf

BEIS. (2017e). *Smart Meters Implementation Programme 2017 Progress Update*. London: Department for Business, Energy and Industrial Strategy. https://assets.publishing.service.gov.uk/government/uploads/system/uploads/attachment_data/file/671930/Smart_Meters_2017_update.pdf

BEIS. (2018). *Capacity Market and Emissions Performance Standard Review*. London: Department for Business, Energy and Industrial Strategy. https://assets.publishing.service.gov.uk/government/uploads/system/uploads/attachment_data/file/732546/CM_Review_call_for_evidence_final_4.pdf

Bourne. (2017). *Speech on Smart Metering by Lord Bourne*. London: Department of Energy and Climate Change. https://www.gov.uk/government/speeches/lord-bournes-speech-on-the-governments-vision-for-smart-metering

Boyle, G. (Ed.). (2008). *Renewable Electricity and the Grid*. London: Earthscan.

Brattle. (2017) *Advancing Past "Baseload" to a Flexible Grid*. Brattle Group Paper for the NRDC. http://files.brattle.com/system/publications/pdfs/000/005/456/original/advancing_past_baseload_to_a_flexible_grid.pdf

Buchanan, B. (2015, May 14). When Amateurs Do the Job of a Professional, the Result Is Smart Grids Secured by Dumb Crypto. *The Conversation*. https://theconversation.com/when-amateurs-do-the-job-of-a-professional-the-result-is-smart-grids-secured-by-dumb-crypto-41769?

CCC. (2011, December 20). *Open Letter from the Committee on Climate Change to the Secretary of State*. https://www.theccc.org.uk/wp-content/uploads/2013/02/green-deal-letter-201211.pdf

CCC. (2018). *Reducing UK Emissions – 2018 Progress Report to Parliament*. London: Committee on Climate Change. www.theccc.org.uk/publication/reducing-uk-emissions-2018-progress-report-to-parliament/

Clegg, N. (2012, April 11). Green Economy Speech. *The Guardian*. https://www.theguardian.com/environment/2012/apr/11/nick-clegg-speech-green-economy

CLG. (2010). *Code for Sustainable Homes, Technical Guide*. London: Department for Communities and Local Government. https://assets.publishing.service.

gov.uk/government/uploads/system/uploads/attachment_data/file/5976/code_for_sustainable_homes_techguide.pdf

CMA. (2018). *CMA Considers Regulation for Heat Networks*. London: Competition and Markets Authority. http://www.gov.uk/government/news/cma-considers-regulation-for-heat-networks

Coyne, B. (2018, February 1). Capacity Market Clears at £6. *The Energyst*. https://theenergyst.com/capacity-market-clears-under-10/

Cuthbertson, A. (2017, November 11). Bitcoin Mining on Track to Consume All of the World's Energy by 2020. *Newsweek*. http://www.newsweek.com/bitcoin-mining-track-consume-worlds-energy-2020-744036

Dawson, A. (2018). *Small Modular Nuclear: Crushed at Birth* (Global Warming Policy Foundation Briefing Paper 33). https://www.thegwpf.org/content/uploads/2018/09/Dawson2.pdf

DCLG. (2013). *Next Steps to Zero Carbon Homes – Allowable Solutions*. London: Consultation, Department for Communities and Local Government. https://assets.publishing.service.gov.uk/government/uploads/system/uploads/attachment_data/file/226610/130731_ALLOWABLE_SOLUTIONS_CONDOC_FOR_PUBLISHING.pdf

DCLG. (2014). *Next Steps to Zero Carbon Homes – Allowable Solutions*. Government Response and Summary of Responses to the Consultation. London: Department for Communities and Local Government. https://assets.publishing.service.gov.uk/government/uploads/system/uploads/attachment_data/file/327842/140626_Government_Response_to_Consultation_-_Next_Steps_to_Zero_Carbon_H__FINAL.pdf

DECC. (2010a). *National Renewable Energy Action Plan for the UK*. London: Department of Energy and Climate Change. https://assets.publishing.service.gov.uk/government/uploads/system/uploads/attachment_data/file/47871/25-nat-ren-energy-action-plan.pdf

DECC. (2010b). *No Place for Cowboys in Green Deal*. London: Department of Energy and Climate Change. https://www.gov.uk/government/news/no-place-for-cowboys-in-green-deal

DECC. (2012). *Electricity Demand Reduction*. London: Department of Energy and Climate Change. http://www.decc.gov.uk/en/content/cms/emissions/edr/edr.aspx

DECC. (2013a). *Smart Metering Implementation Programme*. London: Department of Energy and Climate Change. https://assets.publishing.service.gov.uk/government/uploads/system/uploads/attachment_data/file/87894/ND17Jan_SMIP.pdf

DECC. (2013b). *Local Authorities Leading the Way on the Green Deal*, Green Deal and ECO Team. London: Department of Energy and Climate Change. http://webarchive.nationalarchives.gov.uk/20140403095554/; http://blog.decc.gov.uk/2013/09/12/local-authorities-leading-the-way-on-the-green-deal/

DECC. (2014a). *3rd Annual Progress Report on the Roll-Out of Smart Meters*. London: Department of Energy and Climate Change. https://www.gov.uk/government/collections/annual-progress-report-on-the-roll-out-of-smart-meters

DECC. (2014b). *Research into the Green Deal and ECO Programme Supply Chain*. London: Department of Energy and Climate Change. http://www.gov.uk/government/publications/research-into-the-green-deal-and-eco-programme-supply-chain

DECC. (2014c). *Green Deal Customer Journey Survey: Summary Report – Quantitative Survey Wave 3*. London: Department of Energy and Climate Change. https://www.gov.uk/government/publications/green-deal-customer-journey-survey-summary-report-quantitative-survey-wave-3

DECC. (2014d). *Green Deal Home Improvement Fund Reaches £50 Million Milestone in Six Weeks*. London: Department of Energy and Climate Change. https://www.gov.uk/government/news/green-deal-home-improvement-fund-reaches-50-million-milestone-in-six-weeks

DECC. (2014e). *Delivering UK Energy Investment: Networks 2014*. London: Department of Energy and Climate Change. http://www.gov.uk/government/publications/delivering-uk-energy-investment-networks-2014

DECC. (2015a). *Government Response to the House of Lords Science and Technology Select Committee Inquiry: The Resilience of the Electricity System*. London: Department of Energy and Climate Change. http://www.gov.uk/government/publications/government-response-to-the-house-of-lords-science-and-technology-select-committee-inquirythe-resilience-of-the-electricity-system

DECC. (2015b). *Securing Future Electricity Supply*, Second Capacity Market Auction Results. London: Department of Energy and Climate Change. https://www.gov.uk/government/news/securing-future-electricity-supply

DECC. (2015c). *Green Deal Finance Company Funding to End*. London: Department of Energy and Climate Change. https://www.gov.uk/government/news/green-deal-finance-company-funding-to-end

DECC. (2015d). *£7m Boost to Heat Industry Innovation*. London: Department of Energy and Climate Change. https://www.gov.uk/government/news/7m-boost-to-heat-industry-innovation

DECC. (2015e). *Delivering UK Energy Investment: Networks*. London: Department of Energy and Climate Change. https://assets.publishing.service.gov.uk/government/uploads/system/uploads/attachment_data/file/394509/DECC_Energy_Investment_Report_WEB.pdf

DEFRA. (2005). *Biomass Task Force, Report to the Government*. London: Department for Environment, Food and Rural Affairs. Available at: https://www.iea.org/media/pams/uk/PAMs_UKBiomassTaskForce2005.pdf

DENA. (2018). *The Power to Gas Strategy Platform*. German Energy Agency DENA. http://www.powertogas.info/english

Di Christopher, T. (2017, December 21). No, Bitcoin Isn't Likely to Consume All the World's Electricity in 2020. *CNBC*. https://www.cnbc.com/2017/12/21/no-bitcoin-is-likely-not-going-to-consume-all-the-worlds-energy-in-2020.html

DoT. (2018a). *Office for Low Emission Vehicles Web Site*. Department of Transport/BEIS. https://www.gov.uk/government/organisations/office-for-low-emission-vehicles

DoT. (2018b, March 26). *New Greener Police Cars to Run on Hydrogen*. Department of Transport Press Release. https://www.gov.uk/government/news/new-greener-police-cars-to-run-on-hydrogen

Dougan, T. (1998). *Sustainable Housing*. NATTA Report, Milton Keynes.

Douglas, J. (2015). *Heat Insight: Smart Systems and Heat-Decarbonising Heat for UK Homes*. Birmingham: Energy Technologies Institute. http://www.eti.co.uk/wp-content/uploads/2015/03/Smart-Systems-and-Heat-Decarbonising-Heat-for-UK-Homes-.pdf

DTI. (2001). *The Government's Expenditure Plans 2001–02 to 2003–04 and Main Estimates 2001–02*. London: Department of Trade and Industry. http://www.dti.gov.uk/expenditureplan/expenditure2001/objective_c/chapter9/expend.htm

DUKES. (2018). CHP Data from the Annual BEIS Digests of UK Energy Statistics. https://assets.publishing.service.gov.uk/government/uploads/system/uploads/attachment_data/file/729356/Ch7.pdf

EAC. (2015). *The Future of the Green Investment Bank*. Environmental Audit Committee, House of Commons. https://publications.parliament.uk/pa/cm201516/cmselect/cmenvaud/536/53602.htm

Economist. (2016, August 4). Hinkley Pointless. *The Economist*. http://www.economist.com/news/leaders/21703367-britain-should-cancel-its-nuclear-white-elephant-and-spend-billions-making-renewables

Economist. (2017, April 1). Westinghouse Files for Bankruptcy: The Global Nuclear-Power Industry Is Beset with Problems. *The Economist.* https://www.economist.com/business/2017/04/01/westinghouse-files-for-bankruptcy

Edie. (2017, December 4). Autonomous Network Project to Facilitate UK Electric Vehicle Demand. *Edie.net.* http://www.edie.net/news/8/UK-Power-Networks-launches-autonomous-project-to-facilitate-electric-vehicle-demand/

Edie. (2018a, September 5). Smart Technology Could Reduce Annual Cost of Decarbonising Homes by £6.9bn. *Edie.net.* https://www.edie.net/news/6/Report%2D%2DSmart-technology-could-reduce-annual-cost-of-decarbonising-homes-by%2D%2D6-9bn/

Edie. (2018b, August 23). UK Power Networks Puts Forward 'Flexibility First' Vision. *Edie.net.* http://www.edie.net/news/10/UK-Power-Networks-proposes%2D%2Dflexibility-first%2D%2Dapproach-to-lower-renewable-power-costs/

Element Energy and E4Tech. (2018). *Cost Analysis of Future Heat Infrastructure Options.* Report for National Infrastructure Commission, Element Energy and E4Tech. http://www.nic.org.uk/wp-content/uploads/Element-Energy-and-E4techCost-analysis-of-future-heat-infrastructure-Final.pdf

Elliott, D. (2004). Energy Efficiency and Renewables. *Energy and Environment, 15*(6), 1099–1105.

Elliott, D. (2007). A Sustainable Future for Energy?. In D. Elliott (Ed.), *Sustainable Energy.* Basingstoke: Palgrave. Online at https://link.springer.com/book/10.1057%2F9780230378384

Elliott, D. (2012, March). Emergence of European Supergrids. *Energy Strategy Reviews, 1*(3), 171–173. http://www.sciencedirect.com/science/article/pii/S2211467X12000120

Elliott, D. (2015). Energy Use in Buildings. *International Journal of Ambient Energy, 36*(2), 49. http://www.tandfonline.com/doi/full/10.1080/01430750.2015.1013007#abstract

Elliott, D. (2016a). *Balancing Green Power.* Bristol: Institute of Physics Publications.

Elliott, D. (2016b, January). A Balancing Act for Renewables. *Nature Energy, 1*(1). http://www.nature.com/articles/nenergy20153

Elliott, D. (2017a). *Energy Storage Systems.* Bristol: Institute of Physics Publications. http://iopscience.iop.org/book/978-0-7503-1531-9

Elliott, D. (2017b). *Nuclear Power: Past, Present and Future.* Bristol: Institute of Physics Publications.

Energy and Climate Committee. (2016, September 9). *Government to Miss 2020 Renewable Energy Targets*. Energy and Climate Change Committee. https://www.parliament.uk/business/committees/committees-a-z/commons-select/energy-and-climate-change-committee/news-parliament-2015/heat-transport-report-published-16-17/

Energy Manager. (2018, February 9). Capacity Market Auction T4 Results. *Nationwide Utilities*. http://www.nationwideutilities.com/resources/2018/capacity-market-t4-results/

Energyst. (2016). The Heat Report. *The Energyst*. http://theenergyst.com/wp-content/uploads/2016/04/The-Heat-Report-1.pdf

ETI. (2017, November 5). *Savings of Up to £30 Billion Could Be Realised for UK Heat Networks*. Loughborough: Energy Technologies Institute. http://www.eti.co.uk/news/savings-of-up-to-30-billion-could-be-realised-for-uk-heat-networks

Evan, S. (2017, July 13). Factcheck: How Much Power Will UK Electric Vehicles Need? *Carbon Brief*. https://www.carbonbrief.org/factcheck-how-much-power-will-uk-electric-vehicles-need

FES. (2018). *Future Energy Scenarios*. National Grid, 2018 Update. http://fes.nationalgrid.com/media/1363/fes-interactive-version-final.pdf

FISES. (2016). *Supergrid Study: Approach for the Integration of Renewable Energy in Europe and North Africa*. Werner Platzer Fraunhofer Institute for Solar Energy Systems ISE. https://www.ise.fraunhofer.de/content/dam/ise/en/documents/publications/studies/Study_Supergrid_final_160412_.pdf

Gardiner, J. (2015, August 7). The Green Deal: A Catastrophe from Start to Finish. *Building*. https://www.building.co.uk/focus/the-green-deal-a-catastrophe-from-start-to-finish-/5076906.article

GIB Accounts. (2012–16). Green Investment Bank Annual Reports and Accounts Are at: https://www.gov.uk/government/publications?departments%5B%5D=uk-green-investment-bank

Glover, T. (2016). *Green Gas book, ENA /Parliamentary Labour Party Energy and Climate Change Committee*. https://alansenergyblog.files.wordpress.com/2016/07/final-the-green-gas-book_96pp_v5.pdf

Goodall, C. (2017, July 5th). Hydrogen Made By the Electrolysis of Water Is Now Cost-Competitive. *Carbon Commentary*. http://www.carboncommentary.com/blog/2017/7/5/hydrogen-made-by-the-electrolysis-of-water-is-now-cost-competitive-and-gives-us-another-building-block-for-the-low-carbon-economy

Gosden, E. (2013, December 26). Energy Bill-Payers Face Hundreds of Millions of Pound Charges for 'Pointless' Smart Meter Displays. *Daily Telegraph*.

https://www.telegraph.co.uk/finance/newsbysector/energy/10526915/Energy-bill-payers-face-hundreds-of-millions-of-pound-charges-for-pointless-smart-meter-displays.html

Gosden, E. (2014a, December 19). Households to Pay New £11 Energy Bill Levy to Keep the Lights On. *Daily Telegraph*. http://www.telegraph.co.uk/finance/newsbysector/energy/11303088/Households-to-pay-new-11-energy-bill-levy-to-keep-the-lights-on.html

Gosden, E. (2014b, July 24). Green Deal Cashback Scheme Shuts With Immediate Effect as Homeowners Exhaust £120m Fund. *Daily Telegraph*. http://www.telegraph.co.uk/finance/personalfinance/energy-bills/10989815/Green-Deal-cashback-scheme-shuts-with-immediate-effect-as-homeowners-exhaust-120m-fund.html

Gosden, E. (2016, September 9). Government's Green Heating Plans Branded a 'Total Waste of Money'. *Daily Telegraph*. https://www.telegraph.co.uk/business/2016/09/08/governments-green-heating-plans-branded-a-total-waste-of-money/

Greene, L. (2017, November 15). It's Time to Shake Off Solar Thermal's Preconceptions. *Solar Power Portal*. https://www.solarpowerportal.co.uk/blogs/its_time_to_shake_off_solar_thermals_preconceptions

Grimwood, T. (2018, February 14). Capacity Market: Interconnectors Beat New Gas. *Utility Week*. https://utilityweek.co.uk/capacity-market-interconnectors-beat-new-gas/

Gross, R., Heptonstall, P., Anderson, D., Green, T., Leach, M., & Skea, J. (2006). *The Costs and Impacts of Intermittency*. London: UK Energy Research Centre. http://www.ukerc.ac.uk/publications/the-costs-and-impacts-of-intermittency.html

Grubb, M. (1991, September). The Integration of Renewable Electricity Sources. *Energy Policy, 19*(7), 670–688. http://www.sciencedirect.com/science/article/B6V2W-48XK3MC-1CN/2/e44ea688ee46fbaa053bb793cd73ad78

Gurzu, A. (2017, August 22). SolarCoin Hopes to Ride the Bitcoin Buzz. *Politico*. https://www.politico.eu/article/solarcoin-hopes-to-ride-the-bitcoin-buzz/

H21. (2016). H21 Leeds City Gate Team Report. Northern Gas Networks. https://www.northerngasnetworks.co.uk/wp-content/uploads/2017/04/H21-Report-Interactive-PDF-July-2016.compressed.pdf

Helm, D. (2018, February 25). Smart Meters. *The Times*. https://www.thetimes.co.uk/article/a-stupid-way-to-organise-a-smart-meter-revolution-lvd2355js

Heptonstall, P., Gross, R., & Steiner, F. (2017). *The Costs and Impacts of Intermittency – 2016 Update*. UK Energy Research Centre Report. http://www.ukerc.ac.uk/news/government-must-act-urgently-on-power-system-flexibility-to-avoid-costs-escalating.html

Herring, H., & Sorrell, S. (Eds.). (2009). *Energy Efficiency and Sustainable Consumption*. Basingstoke: Palgrave.

HMG. (2013). *Smart Metering Implementation Programme: Information Leaflet*. London: UK Government. http://www.gov.uk/government/publications/smart-metering-implementation-programme-information-leaflet

HMG. (2018). *The Road to Zero*. HM Government Industrial Strategy Transport Report. https://assets.publishing.service.gov.uk/government/uploads/system/uploads/attachment_data/file/724391/road-to-zero.pdf

Holliday, S. (2015). 'Base-Load Is Outdated'-National Grid CEO Says. *Energy Post*. http://www.energypost.eu/interview-steve-holliday-ceo-national-grid-idea-large-power-stations-baseload-power-outdated/

Holmstedt Pell, E. (2016, June 16). Smart Meters 'Not Needed' After All for European Power Grid. *Euractiv.com*. https://www.euractiv.com/section/energy/news/smart-meters-not-needed-after-all-for-european-power-grid/

Huhne, C. (2012, April 17). Quoted in Shipman, T., & Walker, K. Conservatory Tax? *Daily Mail*. https://www.pressreader.com/uk/daily-mail/20120417/281891590264052

Hunn, N. (2013, December 18). *Smart Metering Is FCUKED, A Disaster*. Claverton Energy Research Group. https://claverton-energy.com/smart-metering-is-fcuked-a-disaster.html

ICL. (2018). *Analysis of Alternative UK Heat Decarbonisation Pathways*. Imperial College London Report for the Committee on Climate Change. http://www.theccc.org.uk/wp-content/uploads/2018/06/Imperial-College-2018-Analysis-of-Alternative-UK-Heat-Decarbonisation-Pathways-Executive-Summary.pdf

Innovate UK. (2018). *Design and Trial Smart Energy Systems: Apply for Funding*. Innovate UK/ UK Research and Innovation. http://www.gov.uk/government/news/design-and-trial-smart-energy-systems-apply-for-funding

IRENA. (2015). *Renewables and Electricity Storage Report*. Abu Dhabi: International Renewable Energy Agency. http://www.irena.org/menu/index.aspx?mnu=Subcat&PriMenuID=36&CatID=141&SubcatID=600

ITV. (2017, February 2). Think Your Smart Meter Is Free? Think Again. *ITV News*. http://www.itv.com/news/2017-02-02/smart-meters-what-you-need-to-know/

Jones, D. (2014, December 19). Quoted in Hope, M. Capacity Market Secures Some New Gas While Providing Stay of Execution to Old Coal. *Carbon Brief*. http://www.carbonbrief.org/blog/2014/12/capacity-market-secures-some-new-gas-while-providing-stay-of-execution-to-old-coal-%281%29/

Labour. (2017). Green Gas book, ENA/Parliamentary Labour Party Energy and Climate Change Committee. https://alansenergyblog.files.wordpress.com/2016/07/final-the-green-gas-book_96pp_v5.pdf

Lloyd, H. (Ed.) (2018). *A Distributed Energy Future for the UK*. London: Institute for Public Policy Research. https://www.ippr.org/files/2018-09/a-distributed-energy-future-for-the-uk-september18.pdf

Lockwood, M. (2018, January 15). *The Capacity Market – How Did We Get Here?* IGov. Exeter University. http://projects.exeter.ac.uk/igov/new-thinking-the-capacity-market-how-did-we-get-here/

Lords. (2008). *The Economics of Renewable Energy*. House of Lords Select Committee on Economic Affairs, Fourth Report, Session 2007–8, November, para 105. http://www.publications.parliament.uk/pa/ld200708/ldselect/ldeconaf/195/19502.htm

MacLean, K., Sansom, R., Watson, T., & Gross, R. (2016). *Managing Heat System Decarbonisation*. Imperial College Centre for Energy Policy and Technology. http://provpol.com/wp-content/uploads/2016/05/Heat-infrastructure-paper.pdf

Mark, L. (2016, May 16). Zero Carbon Homes Policy Scrapped – Again. *The Architects Journal*. https://www.architectsjournal.co.uk/news/zero-carbon-homes-policy-scrapped-again/10006475.article

Meadows, S. (2017, September 8). Is My Smart Meter Charging Me for the Energy Generated by My Solar Panels? *Daily Telegraph*. https://www.telegraph.co.uk/bills-and-utilities/renewable-energy/smart-meter-charging-energy-generated-solar-panels/

Means, E. (2017, April 14). *UK Smart Meters – a Ghastly Mess Created by DECC*. Energy Matters. http://euanmearns.com/uk-smart-meters-a-ghastly-mess-

Milborrow, D. (2007). Wind Power and Similar Renewable Sources – Why Variability Doesn't Matter. In D. Elliott (Ed.), *Sustainable Energy*. Basingstoke: Palgrave. Online at https://link.springer.com/book/10.1057%2F9780230378384

Mitchell, C. (2014, December 19). Capacity Market 'Wasteful, Thoughtless and Parochial'. *ECIU.net*. https://eciu.net/press-releases/2014/capacity-market-wasteful-thoughtless-and-parochial

Morley, K. (2017, March 6). Smart Energy Meters Giving Readings Up to Six Times Too High, Study Finds. *Daily Telegraph*. https://www.telegraph.co.uk/

news/2017/03/06/smart-energy-meters-giving-readings-seven-times-high-study-finds/

Morris, C. (2018). *Can Reactors React?* Potsdam: Institute for Advanced Sustainability Studies (IASS). http://publications.iass-potsdam.de/pubman/item/escidoc:2949898:4/component/escidoc:2949901/IASS_Discussion_Paper_2949898.pdf

MPS. (2003, December 22). Regenesys Runs Out of Time. *Modern Power System.* http://www.modernpowersystems.com/news/newsregenesys-runs-out-of-time

Musaddique, S. (2017, September 6). Wind Turbines Could Power Scottish Smelter and Steel Plants. *The Independent.* http://www.independent.co.uk/news/business/news/wind-turbines-scotland-smelter-steel-plants-mills-highlands-gfc-alliance-motherwell-a7932946.html

NAO. (2016). *Green Deal and Energy Company Obligation*. London: National Audit Office. https://www.nao.org.uk/wp-content/uploads/2016/04/Green-Deal-and-Energy-Company-Obligation.pdf

NAO. (2017). *The Green Investment Bank*. London: National Audit Office. https://www.nao.org.uk/wp-content/uploads/2017/12/The-Green-Investment-Bank-Summary.pdf

NAO. (2018). *Low Carbon Heating of Homes and Businesses and the Renewable Heat Incentive.* London: National Audit Office. https://www.nao.org.uk/report/low-carbon-heating-of-homes-and-businesses-and-the-renewable-heat-incentive/

National Grid. (2014). *T4 Capacity Market Provisional Action Results*. National Grid. https://www.gov.uk/government/uploads/system/uploads/attachment_data/file/389832/Provisional_Results_Report-Ammendment.pdf

NATTA. (1987). *Solar in the City: Guide to Solar Houses and Energy Projects in Milton Keynes*. Milton Keynes: Network for Alternative Technology and Technology Assessment.

NIC. (2017). *Smart Power*. London: National Infrastructure Commission. http://www.gov.uk/government/uploads/system/uploads/attachment_data/file/505218/IC_Energy_Report_web.pdf

NIC. (2018). National Infrastructure Assessment, Chapter 2. In *Low Cost Low Carbon*. London: National Infrastructure Commission. http://www.nic.org.uk/assessment/national-infrastructure-assessment/low-cost-low-carbon/

No2Nucear. (2018). UK Based *No2Nucear* Newsletter. http://www.no2nuclear-power.org.uk/nuclear-news/

Ofgem. (2018). *ECO and Earlier Energy Efficiency Schemes.* London: Office for Gas and Electricity Markets. https://www.ofgem.gov.uk/environmental-programmes/eco/overview-previous-schemes

Oldfield, P. (2015, July 10). UK Scraps Zero Carbon Homes Plan. *The Guardian.* https://www.theguardian.com/environment/2015/jul/10/uk-scraps-zero-carbon-home-target

Orme, B. (2014, April 1). Guest Post: Understanding the Government's Capacity Market. *Carbon Brief.* http://www.carbonbrief.org/guest-post-understanding-governments-capacity-market

Oswald, J., Raine, M., & Ashraf-Ball, H. (2008, August). Will British Weather Provide Reliable Electricity? *Energy Policy, 36*(8), 3212–3225. http://www.sciencedirect.com/science/article/pii/S0301421508002176

OVO/ICL. (2018). *Blueprint for a Post-carbon Society.* OVO Energy/Imperial College London. https://www.ovoenergy.com/binaries/content/assets/documents/pdfs/newsroom/blueprintforapostcarbonsociety-2018.pdf

PAC. (2016). *Government Must Learn From 'Blinkered' Approach to Energy Efficiency.* Public Accounts Committee, Hose of Commons. http://www.parliament.uk/business/committees/committees-a-z/commons-select/public-accounts-committee/news-parliament-2015/household-energy-efficieny-schemes-report-published-16-17/

PAC. (2018). *Renewable Heat Incentive in Great Britain.* Public Accounts Committee Report, House of Commons. https://www.parliament.uk/business/committees/committees-a-z/commons-select/public-accounts-committee/news-parliament-2017/renewable-heat-incentive-report-published-17-19/

Parliamentary Debate. (2015, October 29). Special Debate on the Green Investment Bank, Westminster Hall, Hansard Column 197 WH. http://www.publications.parliament.uk/pa/cm201516/cmhansrd/cm151029/halltext/151029h0001.htm

Policy Connect. (2015). *Policy for Heat: Transforming the System.* Carbon Connect. http://www.policyconnect.org.uk/cc/research/report-future-heat-series-part-2-policy-heat

POST. (2018). *Small Modular Nuclear Reactors.* Parliamentary Office of Science and Technology, POST Briefing Note 580. https://researchbriefings.parliament.uk/ResearchBriefing/Summary/POST-PN-0580

Poyry. (2016). *Cost and Benefits of GB Interconnection.* Poyry Report for the National Infrastructure Commission, London. https://www.gov.uk/government/uploads/system/uploads/attachment_data/file/505222/080_Poyry_CostsAndBenefitsOfGBInterconnection_v500.pdf

Pratt, D. (2018a, February 20). Energy Suppliers Not Up to 'Impossible Task' of Smart Meter Roll-Out Says WHICH? *Current.* http://www.cleanenergynews.co.uk/news/efficiency/energy-suppliers-not-up-to-impossible-task-of-smart-meter-roll-out-says-whi

Pratt, D (2018b, August 30). Almost 1 Million Smart Meters Lacking Functionality Says BEIS as Deployment Continues to Stall. *Current*. http://www.current-news.co.uk/news/almost-1-million-smart-meters-lacking-functionality-admits-beis-as-deployment-continues-to-stall

REA. (2017). *Renewable Energy Review 2017*. London: Renewable Energy Association. http://www.r-e-a.net/upload/final_low_res_renewable_energy_view_-_review_2017.pdf

REM. (2017, November 17). Sheffield Forge Could Be the First in the UK to be Powered by Biomass and Biogas. *Renewable Energy Magazine*. http://www.renewableenergymagazine.com/biomass/sheffield-forge-could-be-the-first-in-20171117

Renew. (2018). Smart Meters – Confusion & Resistance, Renew 232, March–April, Renew on-line 132, Based on Reports from Claverton Energy Research Group Practitioners, p. 5. https://renewnatta.files.wordpress.com/2014/11/rol-132.pdf

Reuters. (2018, April 30). UK Energy Supplier Centrica to Trial Blockchain Technology. *Reuters*. http://www.reuters.com/article/us-centrica-blockchain/uk-energy-supplier-centrica-to-trial-blockchain-technology-idUSKBN1I117G

Richardson, J. (2018, April 28). Solar Power Trade Using Blockchain Completed in UK. *Clean tecnica.com*. https://cleantechnica.com/2018/04/28/solar-power-trade-using-blockchain-completed-in-uk/

Rose, D. (2014, July 6). Unveiled: New £200 'Smart' Meters Every Household Must Pay for (But May Not Work). *Mail On Line*. http://www.dailymail.co.uk/money/bills/article-2681954/Unveiled-New-200-smart-meters-household-pay-not-work.html

Rosenow, J., Guertler, P., Sorrell, S., & Eyre, N. (2018). The Remaining Potential for Energy Savings in UK Households. *Energy Policy, 121*, 542–552. http://www.sciencedirect.com/science/article/pii/S030142151830421X

Sanders, D., Hart, A., Ravishankar, M., & Brunert, J. (2016). *An Analysis of Electricity System Flexibility for Great Britain*. Carbon Trust/Imperial College. https://assets.publishing.service.gov.uk/government/uploads/system/uploads/attachment_data/file/568982/An_analysis_of_electricity_flexibility_for_Great_Britain.pdf

Sansom, R., & Strbac, G. (2012, September). The Impact of Future Heat Demand Pathways on the Economics of Low Carbon Heating Systems. In *BIEE 9th Academic Conference*, Oxford. http://www.biee.org/downloads/the-impact-of-future-heat-demand-pathways-on-the-economics-of-low-carbon-heating-systems/

Scottish Government. (2017). *Scottish Energy Strategy.* Scottish Government. http://www.gov.scot/Resource/0051/00513466.pdf

Shankleman, J. (2014, December 11). £24m Green Deal Fund Exhausted in a Day. *The Guardian.* https://www.theguardian.com/environment/2014/dec/11/24m-green-deal-fund-exhausted-in-a-day

Shannon, L. (2015, May 16). Just How Smart Are These Energy Meters? Hundreds of Thousands of Households Left in Billing Limbo by Clever Gadgets. *This Is Money.* http://www.thisismoney.co.uk/money/bills/article-3084432/ERROR-Smart-energy-meters-leave-hundreds-thousands-households-billing-limbo.htm

Smith, D. (2018, May 23). Quoted in Edie (2018) Decarbonising UK's Heating System Could Cost £450 Billion. *Edie.net.* https://www.edie.net/news/10/Decarbonising-UK-s-heating-system-could-cost%2D%2D450-billion/

Sonnen. (2018). *Sonnen Virtual Community Battery Grid Link System, Promotional Leaflet.* https://sonnenbatterie.de/en/sonnenCommunity

Sovocool, B., Kivimaa, P., Hielscher, S., & Jenkins, K. (2017). Vulnerability and Resistance in the United Kingdom's Smart Meter Transition. *Energy Policy*, 109, 767–781. https://www.sciencedirect.com/science/article/pii/S0301421517304688

Sparrow, A. (2013, November 21). Did David Cameron Tell Aides to 'Get Rid of All the Green Crap'? *The Guardian.* https://www.theguardian.com/politics/2013/nov/21/did-david-cameron-tell-aides-to-get-rid-of-all-the-green-crap

Spicer, M. (1988, December 12). Parliamentary Answer, House of Commons Debate, Hansard Vol. 143 c483W. https://api.parliament.uk/historic-hansard/written-answers/1988/dec/12/alternative-sources-of-energy#S6CV0143P0_19881212_CWA_732

STA. (2016, March 3). Solar Trade Association Press Release. https://www.solar-trade.org.uk/government-proposal-drop-solar-thermal/

Stoker, L. (2018, February 9). T-4 Auction Clears at Record Low Price as Battery Storage Underwhelms. *Current.* https://www.current-news.co.uk/news/t-4-auction-clears-at-record-low-price-as-battery-storage-underwhelms

Tesla. (2018). Powerwall Battery Details: https://www.tesla.com/en_GB/powerwall

Tovey, A. (2017, November 27). GFG Alliance Reveals Strategy to Increase UK Steel Production to 5m Tonnes. *The Telegraph.* http://www.telegraph.co.uk/business/2017/11/27/gfg-alliance-reveals-strategy-increase-uk-steel-production-5m/

Truby, J. (2018, July 24). Decarbonizing Bitcoin: Law and Policy Choices for Reducing the Energy Consumption of Blockchain Technologies and Digital Currencies. *Energy Research & Social Science*, Online. https://www.sciencedirect.com/science/article/pii/S2214629618301750?via%3Dihub

Ueckerdt, F., Hirth, L., Luderer, G., & Edenhofer, O. (2013, December). System LCOE: What Are the Costs of Variable Renewables? *Energy*, (63), 61–75. http://ssrn.com/abstract=2200572

Vaughan, A. (2017, September 18). Most Britons 'Dislike Prospect of Living Near Mini Nuclear Station'. *The Guardian*. https://www.theguardian.com/business/2017/sep/18/most-britons-dislike-prospect-living-mini-nuclear-station

Vaughan, A. (2018a, January 29). Gas-Fired Plants to Reap Huge Subsidies Despite Uncertain Future. *The Guardian*. http://www.theguardian.com/business/2018/jan/29/gas-fired-plants-to-reap-huge-subsidies-despite-uncertain-future

Vaughan, A. (2018b, May 20). Power Station Boss Warns Against Reliance on Electricity Imports. *The Guardian*. http://www.theguardian.com/business/2018/may/20/drax-power-station-boss-warns-against-reliance-electricity-imports-interconnectors

Vaughan, A. (2018c, March 14). Green Investment Bank Sell-Off Process 'Deeply Regrettable', Say MPs. *The Guardian*. http://www.theguardian.com/environment/2018/mar/14/green-investment-bank-sell-off-mps-macquarie

Weaver, M. (2006, December 6). Brown Pledges to Build 'Zero Carbon' Homes. *The Guardian*. https://www.theguardian.com/environment/2006/dec/06/politics.greenpolitics

Whitlock, R. (2017, February 1). UK Smart Energy Rollout Enjoys 81 Percent Approval Rating. *Renewable Energy Magazine*. http://www.renewableenergymagazine.com/energy_saving/uk-smart-energy-rollout-enjoys-81-percent-20170201

Wilkinson, R. (2018, July 21). Smart Meters to Cut Energy Bills by Just £11 a Year, MPs Say. *The Independent*. https://www.independent.co.uk/news/uk/home-news/smart-meters-energy-bills-saving-uk-customers-mps-project-delays-a8457681.html

Williams, S. (2014, October 23). *Parliamentary Answer*. House of Commons. http://www.parliament.uk/business/publications/written-questions-answers-statements/written-question/Commons/2014-10-23/211768/

WNA. (2017, December 7). UK to Support 'Next-Generation' Nuclear Technology. *World Nuclear News*. http://world-nuclear-news.org/NP-UK-to-support-next-generation-nuclear-technology-07121701.html

WNISR. (2018). *World Nuclear Industry Status Report, 2017*, Independent Annual Survey. http://www.worldnuclearreport.org/The-World-Nuclear-Industry-Status-Report-2017-HTML.html

Yeo, T. (2015, March 4). *Use Smart Technology to Cut Peak Demand Rather Than Coal*. Energy and Climate Change Committee, House of Commons, Press Release. https://www.parliament.uk/business/committees/committees-a-z/commons-select/energy-and-climate-change-committee/news/emr-publication1/

8

The Future *into the 2020s and Beyond*

8.1 The Changing Drivers of Technological Change

The UK has been developing renewables since the 1970s, initially stimulated by the 1973–1974 oil crisis. It has not always been easy. As with all technology development, there are risks and uncertainties. The uncertainties were certainly large during the early years, when renewables were often treated as marginal, or, more subtly, as ostensibly important but in practice unfortunately not living up to expectations, so they could be sidelined, as too costly, risky or unreliable. Indeed, in his detailed review of the period up to 1988, Wilson was quite pessimistic about future progress. He presented the UK renewables programme up to that point as 'an essentially tokenistic gesture by the Government', which, in its ostensible aim of coming up with rapid and extensive technological solutions, was 'characterised mainly by fundamentally mistaken assumptions and the consequently misguided programmes' (Wilson 2010). As we have seen the support systems initially adopted (the NFFO and RO) were certainly problematic, despite attempts to improve them, with one retrospective study suggesting that not all the lessons from experience with these

D. Elliott, *Renewable Energy in the UK*, Energy, Climate and the Environment,
https://doi.org/10.1007/978-3-030-04765-8_8

systems had been learned, so that the achievement of programme goals remained uncertain (Wood and Dow 2011).

Be that as it may, the programme went on, with climate change moving up the agenda, indicating that there were also likely to be problems, risks and uncertainties with not taking action. The UK's overall subsequent direction of travel, as set since the adoption of the Climate Change Act in 2008, has been to press ahead with emission reduction and related technical, social and economic measures. We can argue about the costs and the specific technical choices, and even over whether market-based approaches are the most appropriate, but only a few contrarians have challenged the overall climate policy dynamic.

As we have seen, the climate policy dynamic, as established formally in the UK by the Climate Change Act, has driven a series of market-based initiatives which has supported the development of renewables, so that they have reached the point of supplying nearly 30% of UK electricity, with balancing issues also being addressed by a market-based system. Figure 8.1 shows the state of play in 2017 in overall electricity capacity

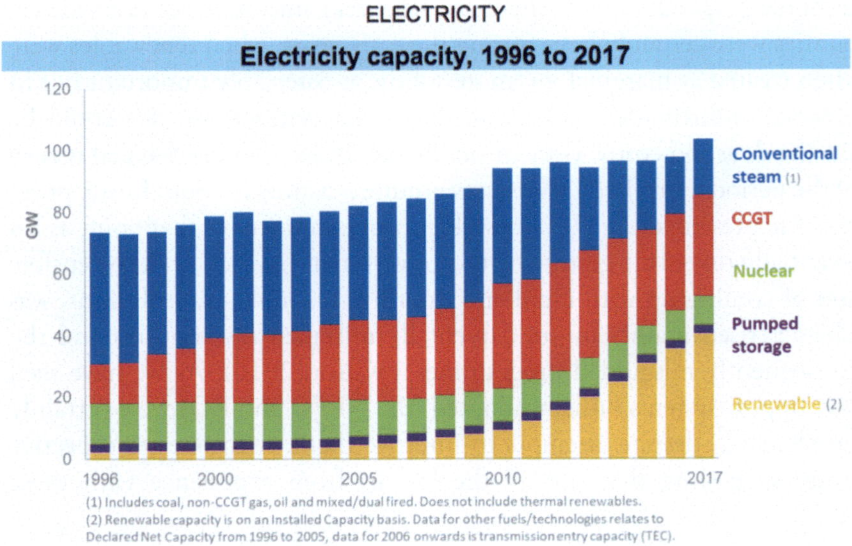

Fig. 8.1 UK Electricity generation capacity in GW. (BEIS 2018a)

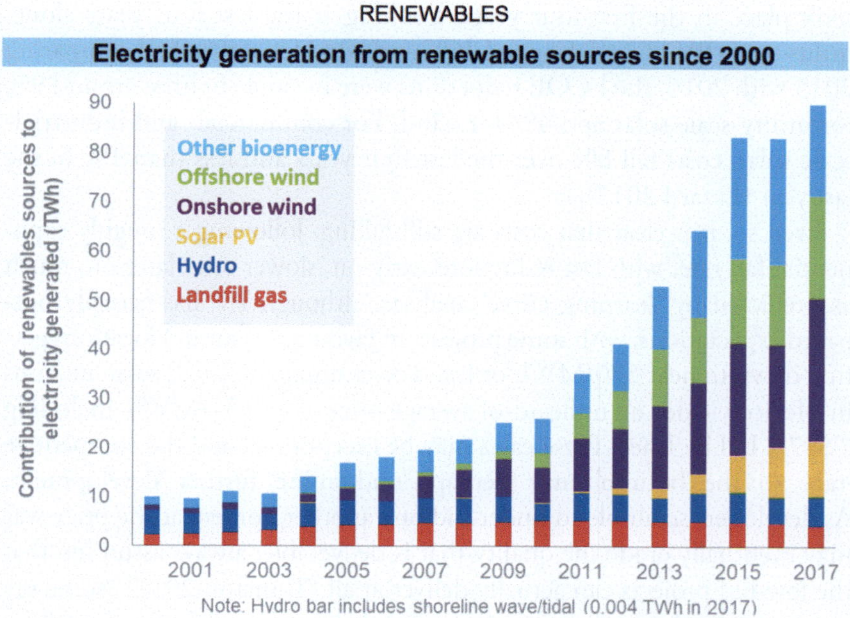

Fig. 8.2 Electricity generation from renewables in TWh. (BEIS 2018a)

terms, while Fig. 8.2 shows the renewable electricity contributions in output terms. We will be looking at the somewhat less impressive results for heat and transport later.

In the next stage, it has been claimed that market forces could take over even more. Certainly, renewable energy supply costs have been falling around the world, so that subsidies might soon not be needed for many of the new supply options. Costs might also fall for the demand management and balancing options, as their use spreads and the technology improves. Moreover, their wide adoption might well reduce overall system costs.

That is of course still speculative: for the moment, balancing still adds costs, perhaps as noted earlier, an extra 10% and possibly more later. However, there have been clear cost reductions on the supply side. Globally, the LCOE for utility-scale solar has dropped 86% in the last eight years. For wind, the figure is 67%, although these high rates have slowed. A study by Lazard consultants said most of these reductions took place in the first four years. Looking at the last four years alone, utility-scale PV LCOE dropped 36% and wind fell 24%. And comparing 2015 with 2016, the LCOE reductions were in single figures: around 9% for utility-scale solar and 4% for wind. For commercial- and industrial-scale solar, costs fell 8% over the last four years and less than 1% in the last year (Lazard 2017).

Even so, it is clear that costs are still falling, following a roughly exponential fall rate, with fast reductions early on, slower ones later on, much as predicted by 'learning curve' analysis, although PV has actually outpaced expectations, with some projects in favourable (sunny) locations getting down to near $20/MWh or less. For example, in 2017, solar auctions in Mexico yielded an unheard-of average price of $20.57/MWh, including a $17.7 bid by Enel. These cases may be exceptional, and the competitive 'race to the bottom' may perhaps undermine further development. As developers scramble to under-bid one another, some fear the price war may eventually erode the quality that is deliverable, always assuming that the low-bid projects can actually deliver at all (Thurston 2017). Some say these winning low-cost projects are speculating against *future* cost falls, a risky strategy (Davis 2017). However, for good or ill, that is how market competition works, and it certainly has led to some cost reductions across the board. As we have seen, that is not just for solar, offshore wind has also exceeded expectations and also looks set to expand globally.

The 'race to the bottom' issue does present potential problems, but they are common in many areas of market competition: all other thing being equal, if prices fall, so do profit margins, and there is less incentive for new investment. However, if overall market demand is expanding, as is the case for most renewables, then there are still commercial gains to be made, so growth can continue. There are of course many issues with growth and with how low costs are achieved. For example, although improvements in the basic cell technology have been one reason for reduced costs of PV, along with the expansion of the market and consequent economies of production scale, at least some of the cost reduction for PV solar modules, and maybe other technologies, is likely to have been due to the low wages and poor working conditions of some of the workforce involved, particularly in parts of Asia. At the very least, there are some key trade union, and indeed ethical, issues to resolve. The energy transition should not be at the cost of worsening exploitation and inequality (Elliott 2015; PCS 2017).

Nevertheless, it might be argued then that, for good or ill, a new dynamic is taking over from the climate response motivation. Renewables are getting so cheap they are being adopted for purely commercial reasons: they are the most economic option in some places. While that may be true in some contexts, at least part of the reason has been earlier support under climate policy-inspired subsidies. Now, arguably, the situation had moved on, and new projects are going ahead in some locations based on their own merits, with, in the UK, there being talk of subsidy-free solar and wind projects. Some examples were mentioned in Sect. 6.4, although, as was noted there, this approach might face limits. Rival energy options might still be subsidised, most obviously nuclear, but also in some countries coal still, while in the UK, shale gas fracking is getting strong government support (as well as facing strong local opposition).

Some renewable energy supporters, reasonably enough, have lamented the lack of a level playing field, although it is perhaps inevitable that emergent technologies have an uphill fight against incumbents. However, some unfair aspects and what are sometimes called 'market failures' can be reduced. The EU's rules limiting 'state aid' are, in part, meant to do this, by blocking support that might create unfair competition, while emission regulations can squeeze out some undesirable options, and, from an optimistic perspective, as subsidies hopefully disappear across the board, with environmental costs being fully reflected in markets, some of the problems faced by emergent clean energy options should reduce.

Nevertheless, quite apart from whatever vested corporate interests might decide to do, with governments still involved with shaping the overall policy context, we will still be left with some potential policy contradictions, as well as some policy failures. The UK energy story has plenty of examples of both, and they bear study since that may prove to be the key to developing better approaches in future. We have looked at many examples of what might be seen as failed or at least perverse policies in the renewable energy field, some of which continue to this day, for example, in relation to the blocks on large PV and onshore wind, and also at the slow and often erratic progress on energy efficiency. There was also the long-running saga on wave energy. In institutional terms, we have seen key agencies and organisations come and go, for good or ill (e.g. ETSU), some as a result of privatisation (e.g. the CEGB) and some due to direct government intervention, for example, the Sustainable Development Commission—with some seeing its abolition in 2011 as reflecting a lack of commitment to long-term sustainability (Monbiot 2010). Nevertheless, the programme continues, with new organisations and new agencies.

This final chapter attempts to take stock of what has emerged from the story so far, and what might happen next, as the climate policy plays out, and as efforts are made to integrate technologies for meeting power, heat and transport energy needs into a balanced and sustainable system, all set within a changing UK political context and changing global context.

8.2 UK Programme Successes, Limits and Policy Influences

Figures 8.1 and 8.2 would seem to indicate that the UK is doing relatively well. However, that has to be put in a global context. The use of new renewable energy technologies is expanding very rapidly around the world, with a capacity growth rate put at around 5.4% over the last ten years, compared to fossil and nuclear which only grew by around 1.6%, while demand has risen by 1.7% (REN21 2018).

In round capacity figures, including hydro, China is in the lead with over 600 GW of renewables installed. The USA is next at near 250 GW. In Germany, wind and solar are at over 100 GW. However, within the EU, in percentage energy output terms, Sweden is the leader. By 2012, it was getting over 50% of its total *energy* (not just electricity) from renewables. Denmark, without the benefits of Sweden's large hydro plants and biomass, was nevertheless at over 25% of all energy by then (it is now at over 30%). Mention should also be made of Norway and Iceland, both outside of the EU, with their very large hydro capacities enabling them to get, respectively, 69% and 73% of their energy from renewables by 2016 (Eurostat 2018).

It is also worth noting that many developing countries also have high overall renewable percentages. For example, many Latin American countries get over 50% of their electricity from hydro, as do some countries in Africa. Indeed, the availability of large hydro inputs means that over 60 countries globally already get the bulk of their electricity from a renewable source. Moreover, as wind and solar are added to the mix, these percentages are rising.

Rapid growth certainly seems likely to continue in most places, with the expectation being that most countries will be able get 50% of their electricity from wind and solar by 2050 (BNEF 2018) and might get to near 100% from all renewables by then (LUT/EWG 2017), with high shares of heat and transport demand also being met using renewable sources. Indeed, it has been claimed that 100% of all energy could be met globally from renewables by 2050 (Jacobson et al. 2017). That is very ambitious, and there have been challenges to that scenario (Clack et al. 2017) and to '100% renewables' scenarios generally (Heard et al. 2017) along with rebuttals (Brown et al. 2018; Diesendorf and Elliston 2018) and debates about how fast renewable energy expansion can occur (Smil 2016; Sovacool 2016). Climate change itself may undermine the prospects for some renewables. Certainly hydro projects in some locations are facing problems with water shortages. However, although the pace may vary around the world, barring major changes or shocks, the transition to very high shares of renewables in most countries seems unstoppable.

Although the UK government often portrays its sustainable energy policies as being exemplary, and its long-term climate/carbon aspirations are certainly ambitious, the sad reality is that, so far, the UK has been trailing a long way behind the leaders in terms of both electricity supply from renewables (see Box 5.2 for the situation up to 2000 and in relation to the 2010 EU Directive electricity targets) and total *energy* supply from renewables. The '15% by 2020' UK renewable energy target agreed with the EU was in any case quite low (one of the lowest in the EU), and it may not actually be reached. Moreover, it may not need to be, if and when the UK leaves the EU. The same may also apply to the newly agreed EU-wide renewable energy target of 32% by 2030 (Darby 2018).

UK progress on meeting the renewable energy targets has certainly been slow. The UK was near the bottom of the EU renewable energy league table, at 4.6%, in 2012, only at that point beating Luxembourg and Malta (Eurostat 2017). That was embarrassing, given that the UK has probably the best renewable resources in the EU, in wind especially. Conservative energy minister Charles Hendry was moved to say 'it is shameful that with some of the strongest winds and highest tidal reaches in Europe, the UK is currently third from bottom in the whole of the EU in its use of renewables' (Hendry 2012). That framing was perhaps unwise for a minister, and, for whatever reason, he was replaced in a reshuffle not long after.

The UK has done a little better since then. By 2015, it was obtaining 8.2% of its energy from renewables. By 2016, it had moved up to 9.3%, but the other EU countries had also increased their share, with, for example, Austria reaching 33.5% in 2016, Latvia 37.2%, Finland nearly 38% and Sweden nearly 54%. So the UK still remained near the bottom of the EU renewable energy league table (Eurostat 2018), at 24th out of 28 in 2016 (Martin 2018).

Some good progress has been made since then on the electricity side (see Figs. 8.1 and 8.2), with the welcome recent boom in offshore wind, along with PV expansion, pushing up the UK's renewables capacity, so that, in 2017, Industry and Climate Minister Nick Hurd could say 'few countries, certainly in Europe, have done more than we have to expand renewable energy electricity capacity since 2010' (Hurd 2017). However, that is a carefully worded claim: it is just for *electricity capacity.* Most other countries in the EU (Germany notably apart) have smaller economies than the UK, with, consequently, less need for generation capacity of any kind. Nevertheless, although the UK's efforts have meant that it now gets around 30% of its electricity *supply* from renewables, others are still doing better in electricity supply percentage terms. For example, at the end of 2016, Denmark and Portugal both got around 54% of their electricity from renewables, Sweden 65% (REN21 2018), and outputs have risen since then. Germany, with a larger economy, a larger power system and a lower wind resource, got over 41% of its electricity from renewables in the first half of 2018, this despite the slow-down from having abandoned the FiT system.

So, to summarise, while the UK may be catching up a little on renewable *electricity* capacity, as noted above, its overall renewable *energy*

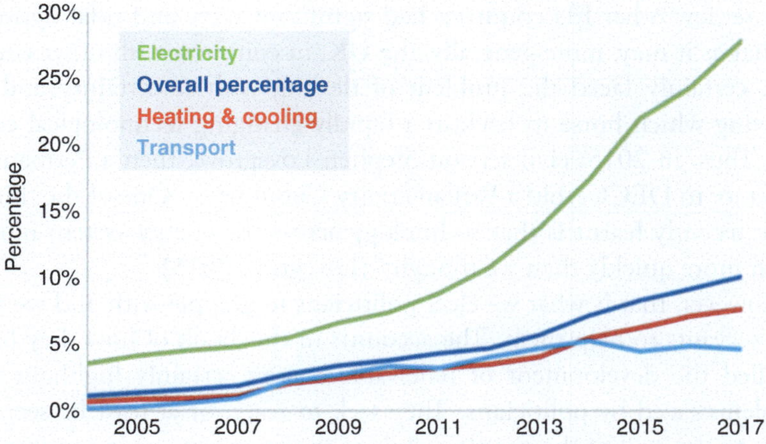

Fig. 8.3 UK progress against 2009 EU renewables energy directive. (BEIS 2018a)

contribution (power, heat and transport) is still very low compared with most of the rest the EU. It had still only reached 10.2% of overall energy by 2017 (BEIS 2018a). As Fig. 8.3 illustrates, transport has been a major problem area for the UK, heat is doing a little better, but even with the boom in green electricity, there is some way to go to meet the overall 15% by 2020 UK renewable energy target. Some other EU countries are also having problems meeting their generally higher targets, with transport being an issue in most places, but overall the UK still lags behind.

We have seen some of the reasons for the UK's relatively poor historic performance and why it may continue to be a laggard. The story can be portrayed just as one of resistance to change, or even of a conspiracy by supporters of the technological status quo. Putting it less aggressively, there do seem to have been some less than ideal support policies, buttressed, at times, by a lack of vision and by commitments to, and the influence of, energy incumbents. That was certainly a contention in some earlier critiques, with, for example, Catherine Mitchell arguing for political and regulatory change (Mitchell 2008).

As we have seen, that view still retains its force. A paradigm shift may be gradually occurring, but the dangers of 'lock in' to existing pathways and technologies still remain. Being charitable, it might be argued that the UK faced particular problems since there were *so many* options to explore: few other EU countries had significant wave and tidal options. Be that as it may, more generally, the UK, in common with other countries, certainly faced the problem of dealing with uncertainty and of knowing which horse to back in a rapidly changing technological context. Thus, in 2015, civil servant Stephen Lovegrove, then a Permanent Secretary to DECC, told a Parliamentary Committee: 'One of the things we've recently learnt is that technology across the energy system moves much more quickly than we thought' (Lovegrove 2015).

However, that is what we elect politicians to grapple with and we pay civil servants to help them. The accounts in this book of how they have handled the development of renewable energy certainly highlight the problems faced by politicians. They seek to react, or at least be seen to react, to the 'will of the people', while at the same time trying to impose their own view on how things should go. Party manifesto commitments are usually powerful drivers, reflecting the Party's view on what would be likely to appeal to voters. So if they are elected, they have a mandate. So it was with the 2015 Conservative manifesto commitment to block further support for onshore wind.

The background rationale for policies like this can however change and also be challenged. In the case of onshore wind, it was in part due to hostility to wind farms, an issue often much to the fore in the right-wing press. But it was also said to be about economics. Thus, at one stage, DECC said that 'with the cost of supplying onshore wind falling, government subsidy is no longer appropriate'. Subsidies themselves are usually an anathema to the right, so it is easy to win approval for cutting them. When launching the cut-back proposals for wind, Energy Secretary Amber Rudd said that 'onshore wind has deployed successfully to date and is an important part of our energy mix', but 'Government support is designed to help technologies to stand on their own two feet, not to encourage a permanent reliance on subsidies. We must continue to take tough judgments about what new projects get subsidies' (Rudd 2015).

It was reasonable for politicians to seek to protect the public purse, especially as costs for wind, and also PV, were falling. However, there are other perspectives. The Energy and Climate Change Select Committees Chair, SNP MP Angus Macneil, said: 'It is deeply worrying that the Secretary of State has said that onshore wind will not be able to access the

Government's guaranteed price Contracts for Difference. Onshore wind remains the cheapest form of clean power and consumers may get less new capacity installed for their money if energy levies are used to fund more expensive forms of low carbon power. It is particularly concerning given that the Department admitted in the same hearing that on our present path there could be a shortfall in meeting the fourth carbon budget in the 2020s' (Agora 2015).

Dr Gordon Edge, from the RenewableUK trade lobby, put it more forcefully: 'If we want to keep all our fuel bills as low as possible, the Government shouldn't be trying its damnedest to kill off the onshore wind industry. The CCC says the current low cost of onshore wind at £80/MWh shows it can be subsidy-free by 2020, matching the cost of new gas. Why on earth would any Government want to stop that happening?' (RUK 2015).

However, there were signs of change. Conservative MP Tim Yeo MP, a previous chair of the Energy and Climate Change Select Committee, who in the past had been strongly opposed to onshore wind, had nevertheless recognised that 'there is a cost to saying no to onshore wind', and, in 2015, he reportedly told the Prime Minister: 'There are some parts of the UK where they seem to be happy with onshore wind. If people are happy to see some wind turbines in their area, and the cost of subsiding those is significantly lower than the subsidy we are offering for offshore wind, I would urge him to let the local view prevail' (Gosden 2015).

The debate rumbled on. In 2018, the Environmental Audit Committee complaining about what they saw as a lack of direction in energy policy, with the onshore wind (and PV) cut-backs being cited as cases in point (EAC 2018). Energy Minister Claire Perry reiterated the government's view that 'we do not believe that more large-scale onshore wind is right for England', but eventually agreed to let some go ahead on remote Islands in Scotland (Perry 2018). Meanwhile though, a YouGov poll for the *Sunday Times* found that more than 50% of Conservative voters felt the government should either encourage or allow the building of more onshore wind farms (Renews 2018).

You evidently cannot please all the people all the time, in this area or perhaps any other. Clearly, political views can have a major impact, and there are those who are hostile to renewables, or at least uncertain about the ways in which we have been trying to expand them. Some can be quite forthright. For example, former Environment Secretary Owen Paterson, in a speech to Nigel Lawson's contrarian Global Warming Policy Foundation in 2014, weighed in against renewables. He said that 'onshore wind is already at maximum capacity as far as available subsidy is concerned' and the 'paltry supply of onshore wind, nowhere near enough to hit the 2050 target, has devastated landscapes, blighted views, divided communities, killed eagles, carpeted the countryside and the very wilderness that the "green blob" claims to love, with new access tracks cut deep into peat, boosted production of carbon-intensive cement, and driven up fuel poverty, while richly rewarding landowners'. He also claimed that offshore wind was proving a failure, 'its gigantic costs requiring more than double the subsidy of onshore wind', with the reason why the UK was the world leader in this technology being that 'no other country is quite so foolish as to plough so much public money into it'. Solar power was 'a non-starter as a significant supplier to the UK grid today and will remain so for as long as our skies are cloudy and our winter nights long'. While biomass was not zero carbon, 'it generates more CO2 per unit of energy even than coal' (Paterson 2014).

The media often relays views like this, even when quite extreme, usually tied to claims that consumer power costs are escalating due to climate and energy policies (Ingham 2014). The reality may be different. Most of the extra costs consumers have faced have arguably been due to fossil fuel supply price rises, with, by 2014, the green energy and climate policies evidently only adding around 7% to a typical domestic bill (DECC 2014a). The cost was expected to rise, as the green energy programme developed, but, since some of the government's energy and climate policies were meant to *save* money, the overall real net cost to consumers of the programme up to 2020 was claimed (by DECC in 2014a) to be likely to be 7% *lower* than the cost that consumers would otherwise have faced.

That sounds convincing, but, as we have seen, the energy-saving programme has in fact not been doing too well. Interestingly, an earlier version of DECC's prices report had put the likely net consumer saving by 2020 at 11% (DECC 2013). So perhaps it is fair enough to challenge arguments like this and projections for the future. It has certainly been a contentious area (Full Fact 2013) and the debate on costs, and on the programme as a whole, continues, renewed every time there is an energy price rise or a new policy development. We can expect further rounds.

Robust views and plain speaking are valuable in any debate, although they have to be within bounds, and subject to fact checking. It is all part of the political process of debate and contestation, both externally, in the media, and internally, within the so-called corridors of power. In the latter context, as we have seen, the various Select Committees have provided continual often very critical oversight and scrutiny, with, for example, in 2003, the Science and Technology Committee complaining that 'the sums invested in public RD&D lack focus and are wholly insufficient in helping the UK meet its renewables targets, in absolute terms and in comparison with the UK's competitors' (Science and Technology Committee 2003).

I have quoted many other similar criticisms from Select Committees, and it is clear that they have often produced some very insightful reports, taking evidence from many key players. However, it has been claimed that, with some exceptions (the Public Accounts Committee for example), Select Committees generally only have limited direct political influence, although it has been suggested that they may have more indirect impacts, keeping ministers on their toes (UCL 2011).

In addition to their Departmental sources, Ministers of course have access to their own sources of analysis, on economics, policy and technology, although there may be a tendency for the views proffered to line up in one direction, in defence of the status quo, as currently perceived. For example, it is notable that all the Chief Scientific Advisors (to the DEn, DECC and BEIS) seem to have been pro-nuclear, including the late Professor David MacKay, whose nicely written book, *Sustainable Energy Without the Hot Air*, was very influential (MacKay 2008). He claimed to be neutral ('arithmetic cannot lie') and certainly delivered a useful corrective to wishful thinking, but some of his analysis of renewables, which, unlike nuclear, he considered were unable to meet our needs, now looks dated (Goodall 2017).

Getting completely independent advice is clearly hard. A recent personal recollection from Stephen Heidari-Robinson, one-time energy and environment adviser to Prime Minister David Cameron, gives some feel for the sort of views that can sometimes be offered. He was very upbeat on the UK's successes: 'It's never been clear to me why green groups and the media spend so much time trashing the government's record, when they should be celebrating the UK's achievements and encouraging other countries to follow suit.' However, he was hostile to the EU's influence and its renewable targets: 'Brexit provides an opportunity to strip away these unhelpful EU renewables targets.' He was also pro-nuclear: 'If the UK government decides not to go ahead with nuclear at all, carbon emissions will increase, as there is no cheaper, base-load option for electricity generation to take its place.' And finally he backed markets against intervention: 'Up front subsidies and the state taking decisions away from the market and individuals has led to perverse outcomes … Government intervention, done in the wrong way, doesn't pick losers: it creates them' (Heidari-Robinson 2016).

All good material for a political debate, with a sharp critique of CCS, also added as a side dish. However, at times, while always hopefully up for a good technical or political argument, responding to some of the less credible and coherent challenges from ideologues and the media can get a little wearying. The likes of the Global Warming Policy Foundation and the arguably misleadingly named Renewable Energy Foundation do tend to churn out contrarian views, with the media often lapping them up. While inputs like this can sometimes be entertaining or even useful, as a part of the debate, some beleaguered renewable energy developers and supporters may get a little tired of dealing with endless invective and dismissive assertions. As well as, seemingly, always having to battle it out with nuclear power.

At times, this can seem a Herculean task. As Dr Bridget Woodman, from IGov, Exeter University, has put it, of late 'pretty much every major policy design has been geared towards creating a perfect environment for Hinkley Point C […] The CfD subsidy is complemented by a suite of other UK taxpayer subsidies and guarantees designed to mitigate investment risks for the French and Chinese investors and to guarantee costs for dealing with nuclear waste or paying compensation for nuclear accidents. Putting all of these subsidies in place has required the UK government to essentially redesign the electricity market over the past few years in an effort to create a situation where investment in a new plant looked attractive' (Woodman 2016).

That may be overstating the case, and certainly there are other perspectives. Although they too may be seen as partisan or biased. It all depends on where you stand. It is the case that nearly all Energy Ministers, of whatever party, have been pro-nuclear, although with some shifts: Labour's Tony Benn moved to an anti-nuclear position once out of office, the Lib Dem's Ed Davey moved the other way, backing nuclear once in office. Labour's Peter Hain was seen as anti-nuclear, but only held the post for six months. The main ministerial resistance to nuclear has come from Ministers in other Departments, such as Michael Meacher, when Minister for the Environment. Robin Cook was another prominent Labour front bench opponent. However, of late, the high cost of Hinkley has begun to shift views, so that they might become more in line with those of the public, with, in a 2018 government public opinion survey, 85% backing renewables and only 38% backing nuclear (BEIS 2018b). So views may change with, in this case, it being the changed economics that may now be the main driver.

It is certainly interesting that the UK National Infrastructure Commission's 2018 National Infrastructure Assessment recommended that the government 'should not agree support for more than one nuclear power station beyond Hinkley Point C before 2025', since renewables were getting cheaper and could prove a safer investment (NIC 2018). The NIC's backup study looked to one possible cost optimal mix, with renewables supplying around 80% of UK power by 2050, assuming a fully developed flexible energy system (Aurora 2018). It remains to be seen how the politicians will react.

8.3 Choosing Technology

Widening the focus beyond Ministerial views and the specific problems and issues faced by politicians, the account above of how renewable energy was handled in the UK highlights the more general point that making technology choices is hard. It has proved very difficult to identify (and agree on) which technologies were going to be viable. Views have sometimes swung between extremes.

Initially, PV solar was dismissed as irrelevant for the UK, but now it may even overtake what at one time was seen as a wild 20 GW by 2020 estimate. Wind power, similarly, was at first seen by some as likely to be marginal, but now is a world beater. It is amusing in this context to look back at the assertion in the UK Department of Energy's 1977 Energy Paper 21, that 'although aerogenerators might be considered economic on certain hill sites... a clear economic case cannot be made for a programme large enough to make a significant contribution to the nations' energy supply' (Department of Energy 1977).

On the other side of the equation, wave power, once seen as major option, along with tidal power, has yet to lift off significantly. Optimism is a fine thing, with, for example, in 2011, the Carbon Trust claiming that 'with targeted innovation, energy generation costs for both wave and tidal stream technologies could reduce to an average of 15 p/kWh by 2025- equivalent to today's cost of offshore wind energy', and also saying that, with continued and targeted innovation, 'the UK's best marine energy sites could generate electricity at costs comparable with nuclear and onshore wind', perhaps as soon as 2025 (Carbon Trust 2011).

That may still turn out to be possible, especially for tidal stream power, but as we have seen, in making such forecasts, there are many uncertainties. The account presented in this book might be seen in terms of a narrative arc, starting from early high expectations for renewables, promoted by enthusiasts, and taken up initially by parts of the establishment, followed by a period when more detailed cost estimates began to undermine support, and then leading to the current period when costs now look attractive and estimates of potential energy contributions are expanding again.

There are of course other possible narratives, more concerned with the role of incumbents in resisting change or focusing on the fate of specific options. However, rather than dwelling too much on the reasons why we have ended up where we are, what will have hopefully come out of this study is a heightened awareness of the problems of making choices, but also of the need to get on with it as best we can, rather than just leaving the status quo in place. We may not be able to learn much from history, and some say that what we can learn is often ignored (Hirsh 2011). But an awareness of past failures and successes may still help us move on more effectively. Indeed, some see failures as a vital part of the learning process, as long as attention is paid to them and they are not used as a reason to halt progress.

In an interesting (and amusing) retrospective on the history of the wave programme, wave energy pioneer Stephen Salter has argued that 'every new technology makes many painful mistakes. Many boilers burst, ships sank, and planes crashed before we got them reliable. The mistakes only become less painful if people learn from them. They will learn only if full details of every mistake are circulated throughout the industry.' He adds, a little bitterly, 'this is certainly not happening now. The requirements for raising private investment require the concealment of expensive disasters in the hope that commercial rivals will repeat the mistakes. Most of the money goes to the most convincing liars' (Salter 2016).

Salter's comments reflect the current era, when short-term cost cutting is evidently all that matters, but should also remind us that, even assuming agreement can be reached on the overall direction of travel, there are issues concerning how to proceed in practical, pragmatic terms. He says 'We must find ways to get the right amount of money to front-line engineers as and when they need it. Over-elaborate rules for tenders and contract management will not stop crooks embezzling public money, but they certainly are too complicated for honest engineers to follow unless they also have a Ph.D. in contract law. Perhaps, we should try fast and flexible agreements for people who have shown that they have earned trust with cruel and unusual punishments if they betray it.'

Salter clearly has had some bad experiences with the current approach, and says that 'politicians and investors want to know the bottom line before making any initial investment and are in a position to enforce their wishes for early figures for the cost of electricity. This is very much like people wanting to know the winning horse before placing a bet.' However, even if short-termism can be avoided, and a more risk-taking entrepreneurial approach adopted, it is still hard to know which options should be followed up and how they should be developed. That applies to wave energy as well as to all else. See Box 8.1.

Box 8.1 Rational Choices: The Wave Energy Debate (Yet Again!)

How can we tell if a new technology will become successful? We tend to use economic viability as a test, but often seem to apply that too soon. As we saw (Box 6.9), that was one conclusion that some made in relation to the UK wave-power programme. Too much was expected too soon. But we cannot wait forever, and some technologies may be 'no hopers', however long we wait. It is probably still too early to say for wave energy, so it is perhaps fortunate that, as we have seen, despite often being pronounced dead, it has regularly been revived. It has been the same for some other renewables. Solar, PV, wind, tidal streams have all at some point been written out of the story as unviable.

Some of course see this in conspiratorial terms. It has, for example, been an article of faith for some that wave energy was marginalised by the nuclear lobby, who allegedly saw it as potential rival. This was even raised in parliament in 1989, with the then Energy Minister Michael Spicer being asked if he accepted 'that a substantial body of opinion believes that dirty tricks by nuclear interests in his Department may have contributed to the rundown of offshore wave power research?' He replied 'I categorically refute the dirty tricks allegation. No Government have done more for the promotion, development and research into renewable energy than this Government. [...] All renewable energy sources are reviewed from time to time, and offshore wave power will certainly be reviewed. Large-scale offshore wave power is not one of the sources that we are currently pursuing as hard as the others' (Spicer 1989).

As we have seen, ACORD's assessment of wave energy has been subject to much debate, and there does seem to be evidence of some errors and perhaps even bias (Elliott 1995; Wilson 2010). However, a problem with conspiracy theories is that they assume a degree of foresight that may not exist. For example, if it were true that the nuclear lobby was worried about rival options (as may well be the case), then, if they did seek to suppress wave energy, they may have chosen the wrong option, since it has shown little sign yet of becoming cheap. By contrast, they should have focused on wind power, or even PV, which have in fact done very well, and indeed are now challenging nuclear. Conspiracy theories can get quite convoluted, but in this case at least, focusing on wave energy would have indicated a belief in the initially dominant view that wave energy was going to be a significant option, thus, perversely, reinforcing that view. When the reality was actually different. Or was it? We may never know: it seems unlikely, in hindsight, but wave energy might have prospered more. It still may and perhaps should. It is a large resource, although it might take time to access it.

Short-termism has its appeal, but so does diversity, spreading risk across a wider range of options. New or improved technologies continue to emerge. They too may get cheap, as PV and wind have done. However, forecasting likely 'best bets' and making choices between options is hard. Indeed, some say it is almost impossible to do rationally and that in reality, mostly we default to (or consciously welcome!) a belief-based decision-making process (Yang 2016). The UK story certainly offers some examples of that.

As we have seen, short-term uncertainty and the need for a longer-term view, set against political pressures for instant successes, faced those involved in the early days of the UK renewables programme. Can we learn any lesson from what was done?

The most obvious first point is that there is always a perceived need for rapid response, and, ideally, for projects to have been successfully completed 'yesterday' or earlier. Urgency is not just a matter of impatience or economic short-termism, early progress (from supported effort) can also be strategically important. In 2002, in the run up to the production of Labour's White Paper on Energy, then Energy Minister Brian Wilson said: 'If we had done more on renewables 20 years ago, we would have a better balance now, and we would not be starting from such a low ebb in renewables' (Wilson 2002).

A second lesson is that you need to persevere. For example, it has often been said that the UK seemed to give up too easily on wind power. Thus, in 2012, Energy and Climate Change Committee chair Tim Yeo said 'In the 1980s the UK squandered the lead it had in wind power development, and now Denmark has a large share of the worldwide market in turbine manufacturing.' Looking at wave and tidal power, he added 'It should be a priority for the government to ensure that the UK remains at the cutting edge of developments in this technology and does not allow our lead to slip' (Black 2012).

That leads on to a third key point, and a familiar one: how do you pick winners? As we have seen, it is not easy, although neither is the alternative of funding a wide range of options to see which turns out to be best. That requires large budgets, even if, at the R&D stage, they may not have to be too large, compared with the next phase of full-scale deployment.

Moreover, it seems clear that, in the innovation game, you do have to take risks: you have to invest despite the uncertainties and to look beyond short-term economic returns. As this book should confirm, the UK's record in relation to that in the renewables field has been mixed: it has of late had some successes, but overall, so far, it has been a little erratic, with, arguably, some poor choices being made, and, given the urgency of the climate issue, all of it painfully slow (Elliott 2017). Can it do better in future?

8.4 Supporting Choices Via the Market

Part of problem has been in the approach to supporting renewables once chosen. Their fate has depended heavily on how governments approached the subsequent development and deployment process. Essentially, it was argued, it had to be driven by market forces. That has been true across most of the political spectrum. As New Labour's Energy Minister, Helen Liddell, put it in 2000, 'we need liberalised, competitive markets that reflect true costs and give consumers real choice. Only then can we hope to see increasing demand and decreasing costs become reality and the possibility of a sustainable future' (Liddell 2000).

Given this regime, as and when they became cost competitive, the new technologies would then be taken up, and the government's support role would end. Thus in 2012, under a Conservative-Lib Dem Government, DECC projected that, by the late 2020s, the Government's role in the electricity market 'will largely be restricted to the setting of high-level objectives for diversity and security of supply' (DECC 2012).

Technology choice would then be mostly done by the market. In 2014, DECC said 'The Government's ambition remains to move to a competitive price discovery process for all technologies as soon as possible. Eventually, we will have a technology neutral auction in place for all low carbon generation.' Presumably that would include nuclear, if that option was retained, though that has not so far been subject to market discipline (DECC 2014b).

The idealised free-market model of course begs the question of how, in reality, the necessary new technologies could be developed to become cost competitive in markets dominated by well-established options, some of which were heavily subsidised. The market was far from being a level playing field, with environmental costs also still not being fully reflected. Given that, from the 1990s onwards, it became apparent that some of the alternative technologies might be needed (and were being used elsewhere) and might become competitive, recourse was made in the UK to various 'market enablement' mechanisms to redress what were sometimes seen as 'market failures', starting with the NFFO, then the RO and now the CfD system. This was often done somewhat grudgingly. The fear was that it might lead to heavy state involvement, along with tight regulation, something free-market enthusiasts wanted to avoid at all costs. That way lay socialism! But if these new environmentally desirable energy options were to progress, the costs of helping them to reach competitiveness had to be met. There were, after all, longer-term economic benefits and indeed incentives, including, as became increasingly clear, the avoidance of major social and economic costs from climate impacts (Stern 2006).

The problem was to devise a system that did not undermine market structures and processes. It is fascinating in this context to look at the UK government's response in 2015 to an EU consultation on energy market design. It was pretty forthright: the UK now had it right, with a competitive market approach via the CfD. Not for us FiTs and the like. See Box 8.2.

Box 8.2 Market Design: The UK Government View (HM Government 2015)

The UK government response to an EU consultation on energy market design was quite single minded: 'The UK has a clear preference for incentivising investment in low-carbon electricity generation, including but not limited to renewables, by means of private finance through revenue support based on metered output covering the expected financing period of the plant. We consider that the advantages in economic and financial terms over capital grants or capacity-based payments are compelling, ensuring that the costs are paid for at the time when benefits are accrued.'

However, long-term contracts were needed, especially for independent suppliers, as was offered under the CfD system: 'Experience in the UK has shown that significant investment does not tend to come forward without long term contracts' which 'help provide investment certainty for new generation capacity – investors generally require assurance that capital and operating costs will be covered by expected revenues, resulting in reliable repayment of debt and the delivery of reasonable returns to equity.'

It was argued that 'Even in a stable, liquid market with good visibility of future demand and capacity, long-term contracts between market participants are necessary for investment to come forward. The only exception is for large, vertically integrated energy companies which are able to manage the risks internally.'

The government's statement went on to dismiss other possible types of support: 'Renewable generation investors generally require assurance of subsidy to top-up market revenues and finance their projects. This can be achieved by a fixed feed-in tariff (essentially a long-term contract with consumers), a premium payment (often converted to a stable revenue stream through an offtake contract) or a variable premium (a long-term contract for difference with consumers). We believe the introduction of the CfD in the UK provides sufficient incentives for investment in renewable energy to come forward at a lower cost of capital and therefore at a lower cost to consumers. By providing support through a variable premium, the CfD gives investors price security throughout the duration of the contract, reducing their cost of capital, whilst protecting consumers from excessive support if wholesale prices rise. Costs are minimised further through competition, which has seen a significant reduction in the level of support provided to generators. Under the CfD, generators are required to sell their output on the market, and are responsible for any imbalance costs.'

However, the Capacity Market was also needed to buttress that and maintain security of supply. That offered an additional role for independent suppliers. Moreover, it was noted that the capacity market could also be open to participants from the EU at some point. The parting message (to the EU) thus seemed to be 'join us'!

The UK government has certainly been consistent in its adherence to the competitive market model, but with some novel interventionist modifications. It backed the EU Emissions Trading System, but, since that was not delivering sufficiently high carbon prices to incentivise much extra support for non-carbon options, it introduced a unilateral UK carbon price support system, funded by a levy on consumer power bills. That approach was also used in terms of direct support for non-carbon options. As we have seen, after an initial grants-based approach, the support schemes that emerged also mostly tried to pass the costs on to consumers, rather than to taxpayers. This is an approach that has sometime been labelled disparagingly as a 'stealth tax'.

To reduce the cost burden on consumers, the new direct support schemes (the NFFO, RO, CfD) have made use of competitive contracting arrangements or other market-based mechanisms to constrain prices. The NFFO and then RO had a mixed success, compared with the guaranteed-price FiTs used in most of the rest of the EU, which were far more successful at getting capacity established and, at least initially, at lower cost. The UK government resisted FiTs, but eventually, under strong lobbying pressure, it introduced one for smaller projects, although it was seen as supplemental to the mainstream competitive market-based support systems (RO and then CfD) for larger projects. The UK FiT system was heavily oversubscribed and was cut back to avoid heavy costs being passed on to consumers. With the consumer cost pass-through rising, FiTs are now being phased out across the EU, and in the UK, in preference for competitive market-based systems, so it might be argued that the UK got it right.

However, as we have seen, the UK's market-based RO and CfD support systems have also had problems, with high costs for the RO and spending caps imposed on the CfD. It is not clear that the CfD will be any better at delivering capacity than FiTs. The RO clearly was not. As we have seen, the FiTs, in Germany especially, seemed better suited to the early stage of deployment of technologies like wind, which were relatively near to competitiveness. They certainly helped wind expand rapidly in Germany, while passing relatively low costs on to consumers, less in fact than the UK market-based RO system.

As was noted earlier (Sect. 6.2), part of the reason for the difference was that, under the RO system, the value of the tradable ROCs given to companies for each MWh of green power supplied was determined by market trading and could not be predicted: so it was hard for companies to get good interest rates from banks when they sought to borrow money to invest in new capacity. The result was that they had to charge consumers more. By contrast, under the FiT system, prices were guaranteed for years ahead at known levels, making the investment climate much more stable, leading to lower consumer prices.

Even so, FiTs may not have been ideal for supporting the take-up of initially higher cost technologies like PV solar, since the cost pass-through to consumers was higher. So, although PV prices have fallen (in part due to the success of FiTs across the EU), as take-up accelerated, the extra consumer cost proved provocative in the UK, and also across the EU, with caps being imposed to limit the impact on consumers. You could of course say that the FiTs had done their job in helping to establish a market for PV. Even if that had cost consumers more, the main cost had been borne by the households that invested in PV. There are some underlying issues of equity here (see Box 8.3), but you could argue that the situation has now moved on. Now that PV is cheaper, all could benefit, and FiTs could be good for PV, as well as for other now cheaper technologies.

Box 8.3 Fairness, FiTs and Growth

Some say that it was wrong to try to accelerate PV via FiTs, since it meant that those who could afford to invest in PV got well reimbursed, in effect by all the other consumers, who had to pay for this innovative push involuntarily. The FiT did tease out capital from those who could afford it, but it was unfair for a rich minority to benefit from what amounted to a tax on a poor majority. However, this was evidently, at least initially, less of an issue in Germany, where the uptake of PV was much wider across the population, with affluence arguably being more widespread than in the UK.

That still begs the 'opportunity cost' issue. Would it have been better to spend this money on, say, wind, since at that stage it was cheaper? Though that in turn begs the question of whether the money would be available. An attraction of PV is that individual consumers can buy it for their homes. Micro turbines apart, that is not an option for wind, but larger wind project is, and was, an option for community groups and wind co-ops. Both routes were adopted in Germany, which is why it now has so much locally owned renewable capacity, both wind and PV.

The FiTs usually have an inbuilt price degression system, with tariff levels designed to reduce year by year to respond to likely cost reductions as the market built and the technology got cheaper. In the event, that process happened much faster than expected, so, in a way, FiTs were a victim of their own success. Drastic cuts had to be imposed to slow down expansion, so as the avoid pass-through cost escalation. It is claimed that the UK CfD system should limit this problem, but that is not certain, and there may be new problems to face. Any competitive market-based support system still has to deal with the larger 'problem of success': as costs fall, the incentive for new investment decreases, so, all other things being equal, expansion will slow (Cornwall Insight 2018).

As noted earlier (Sect. 8.1), this 'race to the bottom' problem may be avoided to some extent if demand for green power expands, as it seems to be doing, but then we move onto another issue. Can this be done without loading up costs on consumers? A central issue for the future may well be, how can we ensure that falling costs compensate for the cost of rapid capacity expansion? At present, a global capacity growth rate of over 5.4% has been achieved for new renewables (REN21 2018), with, by 2016, prices/MWh having fallen by around 50% since 2009 for onshore wind and 80% since 2008 for PV modules (BNEF 2016). Can we do better than that? Will growth accelerate? The various high renewables scenarios suggest that it is possible, although investment rates have fallen recently. Can cost reductions continue, so as to drive growth? BNEF claim that the cost of an average PV plant will fall 71% by 2050 and wind energy by 58% (BNEF 2018). In parallel, there will be system costs to consider, but, in the long term, once the new cleaner, more flexible and efficient energy system has been established, it should cost less to run. However, getting there may impose some extra costs. We (and the planet) all stand to gain, long term, but perhaps the key issue is how can the interim costs and risks be distributed equitably?

The pro-FiT argument is clearly not wining at present. FiTs are being abandoned across the EU. However, arguably, that is more to do with an attempt to throttle back on the rising cost of deploying more renewables. This means that, while the market-based auction system being used may cut costs, it may be at the expense of capacity. Some of the successful low-cost bids for capacity may not actually be able to deliver in reality, they may have bid too low, as we saw had happened in the UK earlier with the NFFO. We shall have to wait to see, for example, in the case of some of low-bid projects in the UK CfD. We are back to the 'race to the bottom' problem discussed earlier (Sect. 8.1), and in Box 8.3, competition may have driven prices too low. In which case no, or fewer, new low-cost projects may emerge. That seems to be the case in the wind-power auctions in Germany at present.

There is also the risk of speculation. Developers may bid low in the hope that costs will fall, so that they can deliver the contracted capacity. However, they may not succeed. When DECC addressed these issues in 2015, it said 'speculative bids are undesirable as they can distort auction outcomes, lead to unsigned contracts, leave budget unspent, and increase allocation risk. This benefits neither auction participants, nor consumers.' But it went on 'the current allocation and CfD contract design include a number of features designed to deter speculative bidding, and encourage development' such as 'demonstrating that the project is in receipt of planning permission and a grid connection agreement; a penalty for non-signature, which excludes projects from taking part in another allocation round for at least 13 months; and, and a potential for a termination of the contract associated with failure to meet the Milestone Stone Delivery Date' (DECC 2015).

8.5 System Scale: Another Choice

The debate on funding systems and renewables expansion programmes continues, with some still defending FiTs, at least for small- to medium-scale projects (EWG 2017). The linked debate on project *scale* also continues. In general, those backing FiTs preferred small-scale projects. That preference has been common in Germany, for example amongst green MPs, with there being a strong commitment to local energy projects. However, the scale issue has not troubled UK ministers too much. Being cynical, you might say that big bold projects have their PR attractions, but so do small ones. MPs get good media coverage when opening a new PV array on a school.

By contrast, as we have seen, scale was a key issue for the early AT movement. Its initial focus in the 1970s was on small-scale technology in a community setting, whereas, as it transpired, the smaller projects supported by FiTs apart, it has usually been larger corporate projects that have dominated, for example with huge offshore wind farms, and even with the smaller projects, the technology is very advanced. There is nothing d-i-y about PV arrays linked to Lithium Ion batteries or hydrogen fuel cells, integrated via smart grid management systems, supplied and run by a large power company. To some extent, this was perhaps inevitable: the technology has improved and corporate interests have adopted it and adapted it to their needs. And it works. Moreover, moving upscale, some examples of local, but large scale, renewable supply for large industrial projects, such as steel and aluminium smelters, were cited in Box 7.2.

That does not mean that we have to go for large scale all the time, it is a matter of horses for courses. But you can go too small, as with some of the micro-wind projects promoted in the early 2000s for individual houses. A study by the Building Research Establishment concluded that, given the typical low wind speeds and low efficiency of small radius machines, 'in many urban areas they are unlikely to pay back either their carbon emissions or the home owner's costs for installation and maintenance' (BRE 2007). A study by the Energy Savings Trust suggested that, instead, 'where possible, communities should be encouraged to work together to deploy the largest possible turbines, as opposed to series of individual installations' (EST 2008).

The simple fact is that the output from wind turbines is proportional to the square of the blade (radius) length and the cube of the wind speed, so a large 1000 kW rated machine on a windy site will produce far more power than 1000 small 1 kW machines in low wind speed locations. Opting for small machines can thus, in some locations, be suboptimal in both economic and environmental terms, using more resources and to less effect and with a higher carbon debt per kWh produced, than if larger units were chosen.

So we have learnt to look for optimals, and realised that small is not always the best, most efficient option. Indeed, although it depends on the location, some would say that large wind farms, with large turbines in high wind speed areas, trump all scales of wind generation, with power delivered by the grid to where it is needed. Similarly, for wave and tidal farms. Except in rare locations, they cannot be easily or effectively deployed on a small-scale local basis.

This logic does not apply to all the renewables options. Although there are investment, installation and operational economies of scale with PV solar, giving large utility-scale projects an economic edge, in output terms, PV modules can be efficient at any scale. They can also deliver power direct to users, with no need for power transmission, unless that is domestic-scale users want to sell any excess and import top-up power via the grid. However, for most of the rest of the renewables, we are talking about grid-integrated system at a variety of scales, including some large units.

Of course, you can go too large, and there are also social and community benefits from smaller-scale projects that have to be set alongside the economies of larger scale. Smaller projects can be more easily subject to local control and ownership and arguably fit better with the more community-scaled decentralised approach and lifestyle promoted by the early AT movement. That way, it is claimed, the power of large corporations can be limited.

However, views on that, and the 'small is beautiful' prescription, may be changing. In 2016, AT pioneer Peter Harper commented that, given the pressing climate problems, we need the most efficient technologies. Although small-scale AT projects, and linked social/lifestyle changes to cut demand, were important, 'it seems unavoidable that 'high-tech' must do the heavy lifting, at least in the short term. Thus two different forms of socio-technical transition need to run in parallel.'

Indeed, he went further and noted that 'for entirely pragmatic reasons' in its Zero Carbon Britain plan, the CAT's proposals 'emphasise technical solutions, and try to avoid the need for lifestyle changes', adding that CAT's 'abandonment of lifestyle change as the key component is a considerable shift'. He says that 'this distances CAT from the 'deep green' parts of the environment movement, who usually cleave to the doctrine that Small is (always) Beautiful. For some it is a betrayal of hard-fought ideals. For others it signifies a welcome return to the real world. Many at CAT would say it is to imply a recognition that we are running out of time' (Harper 2016).

That is an interesting commentary from someone who has been involved with AT and CAT, for decades, although, it may overstate the extent of the CAT's swing away from 'pure' AT, and under-estimate the degree to which AT and environmental ideas have actually spread. In the wider world, and outside the UK especially, the use of small-scale renewable energy systems is now very extensive, with, for example, around 40% of Germany's 100 GW or so of renewable capacity now being locally owned, including by domestic consumers and energy co-ops. Some hope that can expand further, with more local power. Greenpeace has argued that, in principle, there could be a 70%/30% split between small local-scale and large central-scale systems in most places (Greenpeace 2015).

We can argue about the exact ratio. Seventy per cent from local projects globally seems rather optimistic, except perhaps for the longer term.

It will depend on the location and on local motivations (Wills 2016). In some locations, there will be little opportunity for local generation, while in others, it may be felt that the social value of community-scale projects outweighs the technical efficiency benefits of larger schemes. Or vice versa. Then again, some locations will be so well placed, in terms of renewable resources, that they can become net exporters of green power. Indeed, this has been happening in many places in Germany, with the rise of many large, 1000 and more member-strong, local energy co-ops (Debor 2014), although recent policies (e.g. the demise of the FiT system) are making it harder for citizen-led projects. Nevertheless, some look to a future in which 'local power' under local control dominates.

Progress in the UK in this regard has been a somewhat limited. Although private consumer uptake of PV has been extensive, and there are some local energy projects, so far there is only around 249 MW of community-owned renewable energy capacity, with the FiT cuts slowing progress (Community Energy England 2018). Nevertheless, progress is being made, for example by the various municipal energy schemes, and there is much that can be done at all levels, including by community-based schemes (Lloyd 2018).

Certainly, around the world, it does seem that new optimals may emerge, with the rise of localisation as a new political ideal and community entrepreneurship as a new social option. Some look to continued expansion of community-based projects, as part of a global energy system shift (Koirala et al. 2016). Changes like that might also imply changes in attitude and consumption patterns, although that is more speculative. Participants in such schemes are likely to be more environmentally sensitive. However, Harper, looking from a UK perspective, is pessimistic. He says 'Had the long slow cultural changes called for in the 1970s taken hold and become widespread, all could have been achieved with 'alternative' systems. But of course these changes have not spread, and as various global thresholds loom, the responses have to be more rapid, drastic, infrastructural and 'one-size-fits-all', and now 'there is much more concern for whether new systems can be generalised and applied on a wider national scale' (Harper 2016).

Harper seems to suggest that, for the bulk of the population, we may still have a long way to go to make the necessary and urgent changes.

That may be too pessimistic, but if he is right, then, if wider and faster changes are needed, in lifestyle as well as technology, tragically, we may need more shocks to motivate change.

8.6 Choices for the Future

What can we learn from the responses to previous shocks, such as the oil crisis and the rise of concern about climate changes? The accounts above of how specific technologies were developed following these major external shocks offers some interesting insights into how we might move on in future and deal with possible future shocks, opening up some wider issues related to the choices we face for the future. Indeed, some say we are in the midst of a possibly terminal crisis for the current way of doing things, so we need to respond fast (Klein 2014; Heinberg 2017).

At the general societal choice level, given the apparent need for change, in response to the earlier episodes, there was a range of possible approaches. 'More of the same' seemed unviable in both cases, although some looked to minor adjustments. But others were more radical, although with some odd twists. In the 1970s, in the face of the oil crisis, the environmental and AT movement in effect offered revamped technologies from past, 'going back to earlier paths', along with prescriptions for a less consumerist and materialistic life style. Others, more establishments inclined, looked to accelerated reliance on new 'advanced' technologies, like nuclear. Both groups adopted the same responses when, later, they were faced with climate change.

In so doing, the establishment-orientated group seemed to be suggesting that there was only one way forward, based on the logic of technology, which was being revealed, sometimes under pressure, by our efforts, as part of what we have called 'progress', driven primarily by economics. Despite pragmatic adjustments in vision and technological prescription (and Harper's pessimism), the old AT view may still retain some force: there are multiple possible ways forward, each supported by different technologies and values, the main task being to choose which paths and mixes were felt to be best in social and environmental terms. Some might also add political, moral, ethical and spiritual dimensions as well.

This big picture may have been behind much AT thinking, but evidently, it played little direct part in what happened subsequently. Renewable energy has become something of a technical fix, allowing more or less the same sort of consumerist society to continue and expand, although with fewer emissions. The issue of sustainable consumption may have become more urgent, given the rise of climate shocks and pollution scares, but there may be some truth behind Harper's pessimism. Few in the affluent countries are prepared to consider major life style changes, beyond a bit more recycling and bicycling. There have been proposals for 'personal carbon rationing', via the issue of (possibly tradable) carbon credit vouchers set against energy-use limits, but, so far, no government has seen fit to impose anything so seemingly draconian. Tragically, unless other actions are taken, consumers may have dramatic changes forced on them as and when the climate/air quality situation worsens. By contrast, the appeal of technical fixes is obvious, and some argue that they can indeed allow us to continue living in more or less the same way, at least in terms of energy: it does seem likely that, if demand growth can be contained, renewables, coupled with efficiency, can allow us to meet most of our energy needs into the far future, without impacting on the climate or having to make radial social changes.

However, there are also other resources' needs and limits (most obviously water and land) and other environmental impacts than just climate change (including threats to biodiversity). Given rapacious consumerism and the drive to endlessly accelerating economic growth, some believe we have already gone beyond the ecological carrying capacity of the planet. The jury is out on whether we can row back from disaster (Barkham 2018). Some say we can make (and actually also benefit from) voluntary changes to our lifestyle and consumption habits, which will help to avoid or limit ecological crises—energy diets, and perhaps also (less) meat diets! But, even if we do that and more, and also adopt the best ecological energy technology, we may still be approaching the point when some more radical system-wide choices have to be made, assuming it is not already too late. That may be too bleak a view. But if the worst comes to the worst, hopefully we can do better than last time in response to earlier shocks, and hopefully, to aid that, this book will provide at least some help by identifying how things went then. The shocks then may turn out to be relatively minor, compared with what we may soon face, but lessons can be learned from what was done. Moreover, the changes and challenges that were made in response were not all a waste of time. They have opened up new pathways, and we have been able to travel quite some way down them, despite all the resistance. Much further than was once imagined. But even being optimistic, there is still some way to go and maybe new pathways to find.

8.7 Conclusions

We have moved a long way from the alternative technologies promoted in the early years by the AT movement and its allies. Many of the early activists have moved on. Some got involved in grass-roots energy projects, including community-owned projects, and that idea was followed up in more recent projects like the Westmill wind and solar farm near Swindon, championed by environmental activist and organic farmer Adam Twine (Co-op Energy 2018). Others moved on to deploying green energy technologies commercially. Dale Vince, once very much a member of the counter-culture, went on to set up Ecotricity in 1997. It has become one of the UK's main independent renewable energy generation and retail companies. Its success and his iconoclastic quirkiness have made him something of a celebrity in some circles (Flintoff 2011). Jeremy Leggett, once a leading energy campaigner with Greenpeace, went on to set up Solar Century, now one of the main solar PV retail and installation companies in the UK. His account of their history is fascinating (Leggett 2018).

There are many more individuals that could be named, including, as a welcome change, some women, including Juliet Davenport, who set up what became Good Energy, now a leading UK green energy retailer. Interestingly, that approach may spread. Some green power tariffs are now being offered at competitive prices, so it may be that more consumers will get into the so-called voluntary green power retail market, as an easier alternative than the 'prosumer' self-generation route, given the abolition of the FiT (Vaughan 2018).

Direct involvement with generation and marketing has not been the only destination for some of the early activists. Many other individuals, from academia or other such trades, have also played key roles over the years. In the UK and elsewhere, many of the men and women involved with the development of renewable energy in various guises can trace their roots back to the 1970s and to some of the more challenging non-commercial value systems that were prevalent then. However, there is also now new generations coming up, who did not live through that, and are immersed in the modern hyper-competitive world. In part, this book is for them, in an attempt to report on some limited aspects of what the first generation experienced. It is also for outsiders who might like to know about some of the history of why we (and I count myself as one of the players) have done what we have done. But it is also about the future—what happens next?

The story I have told is partly about the loss of innocence: many of the AT ideals have been lost. Nevertheless, some of the dynamism has survived, and the renewables story brings us up to a point where a wide range of alternative technologies are more or less in place, and are now seeking to find their roles in an integrated system. The new variants are more efficient and more economic than much of the old AT, in part as a result of exposure to market competition. So that exposure has not always been a bad thing. That gain has however been at the expense of handing over control of technology development to commercial interests and governments. The result has been mixed. We now have increasing amounts of large-scale corporately controlled and owned renewable energy technology, but also many locally owned smaller-scale projects, with some consumers able to generate their own power, to the extent that the big power companies are beginning to lose control of the power market in some countries. Some see this ownership and control issue as being fundamental, challenging the economic and (even) political status quo. However, the PV solar equipment typically used by self-generating 'prosumers' still has to be made somewhere, so nothing much has changed there. They are just using new products, made in the same way as before.

Nevertheless, some things do change. The newly emerging cleaner, greener, more localised energy system is different from often very polluting centralised systems that went before. Moreover, more subtly, the drivers of the change process may be changing. As the story told in this book has indicated, these technological changes have so far mainly been the result of reactions to external events. However, as we have also seen, the dynamic that these reactions released may actually have changed the driver for change. Renewables may have been adopted as a response to the oil crisis and then to climate change, and more recently, to the air pollution problems of some cities, but the result has been that they have become increasingly competitive and may see off both fossil fuels and nuclear. So, it may no longer be just climate or environment policy that is driving the technology changes.

Does this indicate that, in the end, technology really does have an internal, almost autonomous, developmental logic that will be revealed and brought into play via markets, given the right conditions? A hidden, but almost predetermined, single path ahead? While some AT enthusiasts may have thought that there was one obvious technological path forward, that was probably not what most of the people in the counter-culture or AT movement thought. They looked to many technical and social paths being open and wanted to let diversity, and '1000 flowers', bloom. Diversity is a good strategic principle—it can avoid 'lock in' to one path and technology, make systems more robust against change and spread the risks of failure across a wider range of options.

In the past, we tend to have focused on one of two major energy options, mostly coal, until recently. One of the attractions of renewables is that there are so many different types, most of them operating on different principles, at a variety of scales and with a range of applications in varied contexts. We have some choices. So we are unlikely to end up converging on a more or less single technologically and economically determined path. If we did, that would be an anathema to many of the AT pioneers.

The future will make itself anew, whatever they thought. However, the changes they helped unleash will hopefully add up to more than just a significant change in energy production, but also help sustain a conviction that there is always an alternative: we do not have to stay on tramlines. Moreover, as some of the history related in the book may illustrate, it is often those on the fringe, often active at the grass roots, who keep reminding us that there are always choices. They may sometimes be oddball, eccentric outliers, but in the end, they, along with engineers who can try to turn ideas into reality, may be one of our most valuable resources.

That is not to say that all innovative ideas come from the margins or the grass roots, although even with the advent of giant R&D institutes, it is still often the case that the most original ideas come from outside, from lone inventors and deviant groups looking for new approaches. That model of innovation can of course be overly romanticised, although in the wider area of IT and digital technology, much of the change in the past few decades has been the result of innovations initiated by the classic 'nerds in a garage'. However, there is a big step from that to wider application and deployment. In the case of IT, there were briefly pockets of enthusiast-users, but Macs, PCs and the rest rapidly moved to commercial uptake via the market. In the case of AT, the early enthusiast phase was more protracted: it was harder to break out of that niche, and the ideas that eventually were transferred may not always have been developed in the way that the AT movement would have liked.

As noted in Chap. 1, some theorists portray 'niches' as vitally important for innovation, providing protected microcosms, 'nurseries' in which new ideas can emerge and be tested, and then spread to be adopted in the wider socio-economic regime. Although some ideas clearly did transfer to wider use, the history of AT arguably provides only limited backing for the idea of that process being widely effective. There were significant constraints on easy translation. As innovation theorist Adrian Smith says, even if it has been demonstrated that an alternative kind of sustainable practice is possible, 'niche diffusion requires sufficient common ground for those practices to link with the [wider] regime. Performance criteria in niche and regime need to come into some kind of correspondence—translating what works in the niche into something that also works in the regime' (Smith 2007). That rarely happened in the case of energy alternatives, although with some exceptions, green building technology for example: building designers sometimes picked up on ideas from the fringe.

While some ideas (and people) did make it out, mostly it proved hard to go beyond the niche. Some of course say we should not try, but instead should just expand the niche, so that it eventually becomes the norm. An optimistic interpretation of events would suggest that this is what has now actually happened: not all of it has spread, but renewables at least are now becoming dominant. However, as we have seen, there are limits to what can be, and has been, transferred, and not all of the ideas that have been picked up and promoted, for example micro-wind systems for domestic use, have proved viable. At worst, especially when installed in low wind speed urban environments, they may become just 'green bling', ostentatious 'add ons', with little or no environmental merit.

None of this is to suggest that 'community niches' cannot play an important role in originating, sustaining, testing and demonstrating new ideas, including in the energy field (Smith et al. 2016). There are many modern-day examples of community-level projects providing test beds for new ideas. As noted above, there can also be some clear local benefits from community involvement, including non-energy benefits, such as social cohesion (Brummer 2018).

However, it seems optimistic to believe that small-scale local renewables, run in a community context, will necessarily dominate in the overall energy transition, or that grass-roots initiatives are the only way forward. They have done well in recent years, in part due to the FiT system, in Germany especially, although wider realities and changes have inevitably intervened. FiTs are on the way out (in the UK too), highlighting the point that government policy changes can have large impacts at the local level.

Some may see this in conspiracy theory terms, arguing that the establishment is, as ever, trying to stop new ideas from spreading, so undermining the markets of the established energy utilities. There may be some truth in that, but the wider reality is probably more complex. Change is underway, but it may involve multiple lines of development, not all of it necessarily being coherent or conflict-free. Small-scale community projects clearly have their place and will hopefully spread, as no doubt will 'prosumer' take-up. Indeed, as we have seen, in some scenarios, private prosumer self-generation and distributed power systems generally expand quite extensively (Lloyd 2018). So we may be able to develop a substantial local generation sector, with many possible variants, depending on the local context. However, some of the running will also inevitably be made by larger corporate projects. There is an economic and technical logic to that: small is not always beautiful, and expertise is also a limited resource, as of course is finance.

For good or ill, outside of Utopia, some accommodation has to be made with the reality that markets will inevitably play a key role in whatever happens next. That said, on their own, markets may not be sufficient to drive the change fast enough to respond to the climate crisis. They can also lead to perverse results. Markets are, arguably, good at making some people rich, but do not necessarily lead to rational, much less equitable, development patterns. That may require markets to be contained and constrained.

For the moment, the UK government sees market competition as being the best way to achieve its energy goals. For example, in the context of its smart grid proposals, it has said that 'at the centre of our approach is ensuring effective markets and competition. Markets can allow the best flexible solutions to flourish and help deliver a secure, affordable and clean energy system. We want to see competition that is as far reaching as possible to make sure consumers benefit from a more efficient system' (BEIS 2017). That is all debatable. As we have seen, adherence to the competitive market approach has not always proved effective in the development of UK renewables, and it has also arguably led to some problems for the various balancing and energy-saving initiatives looked at in the last chapter. We ought to be able to do better.

Certainly, one of the many challenges for the future will be to try to ensure that, whatever projects and programmes emerge, they are done right, in socially and environmentally appropriate ways. That may imply the need for more democratic forms of control. Some see municipal projects playing more of a role, alongside community energy co-operatives, while some even look to nationalisation (Corbyn 2016; PCS 2017; Co-operative Party 2018). Clearly, when it comes to politics and technology, it is still hard to avoid the issues of ownership and control, much as some of the early AT pioneers and radicals argued (Elliott and Elliott 1976). But now we are talking about big technology, as well as small-scale local community Schemes. A new world with more decentralised ownership and control? It is one possible future. Although whether any of that will improve on the often ponderous processes of parliamentary and technocratic oversight that we have at present, as explored in this study, remains to be seen.

8.8 Afterword: AT and the Renewables Revolution

This book has explored what happened to the renewable energy ideas that had been pioneered on the fringe in the AT movement. To round off, here is a simple retrospective, looking back, a little wryly, in the context of 'AT' views and likely reactions.

As we have seen, the UK renewables programme has often focused on giant schemes nearly all of which have been abandoned: the 8 GW Severn Tidal Barrage, the 3 MW Orkney wind turbine, the 2 GW wave programme. It also tried to get these technologies, and the others, to move to competiveness too soon. What has worked around the world have been smaller projects, wind farms and PV solar. They may have been naïve, but the 1970s AT activists may have been right: small is beautiful. Though not too small.

Another key issue has been the support for nuclear power. Throughout most of the period, it was getting £100s of millions for R&D, while renewables were getting £10s of millions.

The nuclear option is now being reconsidered, even in the UK. The 1970s AT activist may have been right on that too: nuclear power, no thanks!

The debate now is whether we can get to 100% renewables by 2050. The AT activists would have liked that. But they would have been surprised by the fact that, although the oil crisis and then climate change provided major boosts for renewables, it is now their low cost that is driving them ahead. They are increasingly the technology of market choice.

Changing our energy system will not be easy, it will cost. The new technologies are getting cheaper, but someone has to pay for their development. Governments worry about loading up taxpayers with cost, and have tried to pass them on to consumers, via the NFFO, RO, FiTs and the CfD, the so-called green levies. But that sometimes is resisted. So a big issue for the next phase may be, will the fall in cost of the technology allow for expansion without provoking consumer backlashes? I am not sure what the AT activists would have made of that. Or the idea that it can be left to the market, or the government, to sort out. More likely they would say 'we have to do it ourselves'. And, dare I say, that might be right in theory, although perhaps not always possible in practice.

Finally, there is the issue of economic growth. Renewables are one of few growth areas in many economies at present, and, if we are to shift to a sustainable energy system, that is good news. But some in the AT and environmental movement looked to a steady-state economy, not reliant on endless growth in consumption. They would no doubt be happy to condone growth in renewables in the short term, but would look to a future when a sustainable system had been developed, after which growth could stop, with just maintenance and occasional efficiency upgrades being required. We are a very long way off that in any sector, with, in any case, growth being one of few ways for many of the world's poor to move beyond a subsistence level existence. Maybe the largest issue of all is whether a transition can be achieved globally, that is, both ecologically and socially viable. I am not sure the AT movement, or anyone else, had or has the answer to that, although some have talked of a 'Green New Deal' as an alternative to a 'Degrowth' strategy (Pollin 2018). However, whichever routes forward emerge, the switch-over to renewables must surely be part of the way ahead, and the UK has made a start.

References

Agora. (2015, July 22). *Planning Bids for Renewable Energy Projects Drop By 80%.* Agora Portal for Parliamentary Development. https://www.agora-parl.org/news/uk-exclusive-planning-bids-renewable-energy-projects-drop-80

Aurora. (2018). *Power Sector Modelling: System Cost Impact of Renewables.* Report for the National Infrastructure Commission, Aurora Energy Research. https://www.nic.org.uk/wp-content/uploads/Power-sector-modelling-final-report-1-Aurora-Energy-Research.pdf

Barkham, P. (2018, April 26). We're Doomed: Mayer Hillman on the Climate Reality No One Else Will Dare Mention. *The Guardian.* http://www.theguardian.com/environment/2018/apr/26/were-doomed-mayer-hillman-on-the-climate-reality-no-one-else-will-dare-mention

BEIS. (2017). *Upgrading Our Energy System: Smart Systems and Flexibility Plan.* London: Department for Business, Energy and Industrial Strategy. http://www.gov.uk/government/publications/upgrading-our-energy-system-smart-systems-and-flexibility-plan

BEIS. (2018a). *Energy in Brief*. Summary of UK Energy Statistics for 2017. London: Department for Business, Energy and Industrial Strategy. http://www.gov.uk/government/collections/uk-energy-in-brief

BEIS. (2018b). *Energy and Climate Change Public Attitude Tracker*, Wave 25. London: Department for Business, Energy and Industrial Strategy. https://assets.publishing.service.gov.uk/government/uploads/system/uploads/attachment_data/file/702640/Wave_25_Summary_Report.pdf

Black, R. (2012, Feb 19). Wave and Tidal Power Need Support, Say MPs. *BBC News*. https://www.bbc.co.uk/news/science-environment-17078726

BNEF. (2016, April). *Bloomberg New Energy Finance Summit*. New York. www.bbhub.io/bnef/sites/4/2016/04/BNEF-Summit-Keynote-2016.pdf

BNEF. (2018). *New Energy Outlook 2018*. Bloomberg New Energy Finance. https://about.bnef.com/new-energy-outlook/#toc-download

BRE. (2007). *Micro-Wind Turbines in the Urban Environment*. Garston: Building Research Establishment.

Brown, T., Bischof-Niernz, T., Block, K., Breyer, C., Lund, H., & Mathiesen, B (2018). Response to 'Burden of Proof: A Comprehensive Review of the Feasibility of 100% Renewable-Electricity Systems. *Renewable and Sustainable Energy Reviews, 92*, 834–847. https://www.sciencedirect.com/science/article/pii/S136403211830307#!

Brummer, V. (2018). Community Energy – Benefits and Barriers. *Renewable and Sustainable Energy Reviews, 94*, 187–196. https://www.sciencedirect.com/science/article/pii/S1364032118304507

Carbon Trust. (2011). *Press Release for the Launch of 'Accelerating Marine Energy'*. Carbon Trust, Quoted in Renew193, Sept–Oct, Report at: https://www.carbontrust.com/media/5675/ctc797.pdf

Clack, C., Qvist, S., Apt, J., Bazilian, M., Brandt, A., Caldeira, K., S Davis, S., Diakov, V., Handschy, M., Hines, P., Jaramillo, P., Kammen, D., Long, J., Morgan, G., Reed, A., Sivaram, V., Sweeney, J., Tynan, G., Victor, D., Weyant, J., & Whitacre, J. (2017). Evaluation of a Proposal for Reliable Low-Cost Grid Power with 100% Wind, Water, and Solar. *PNAS, 114*(26), 6722–6727. http://www.pnas.org/content/114/26/6722

Community Energy England. (2018). *Community Energy: State of the Sector 2018*. Community Energy England/Community Energy Wales. https://communityenergyengland.org/files/document/169/1530262460_CEE_StateoftheSectorReportv.1.51.pdf

Co-op Energy. (2018). *Westmill Wind Farm and Westmill Solar Farm*. Co-op Energy. https://www.cooperativeenergy.coop/why-us/community/community-energy-conference/westmill/

Co-operative Party. (2018). *Democratic Public-Ownership for the 21st Century*. London: The Co-operative Party. https://party.coop/ownershipmatters

Corbyn, J. (2016, September 7). Why Labour Is Putting Energy Reform at the Heart of Its Green Agenda. *The Guardian*. http://www.theguardian.com/environment/blog/2016/sep/07/why-labour-is-putting-energy-reform-at-heart-of-its-green-agenda-jeremy-corbyn

Cornwall Insight. (2018). *Wholesale Power Price Cannibalisation*. Cornwall Insight Paper. https://www.cornwall-insight.com/landing-pages/wholesale-power-price-cannibalisation

Darby, M. (2018, June 20). UK Government Refuses to Commit to EU Clean Energy Targets After Brexit. *Climate Home News*. http://www.climatechange-news.com/2018/06/20/uk-government-refuses-commit-eu-clean-energy-targets-brexit/

Davis, L (2017, December 4). Are Mexican Renewables Really This Cheap?. *Energy Institute at Hass Blog*. https://energyathaas.wordpress.com/2017/12/04/are-mexican-renewables-really-this-cheap/

Debor, S. (2014). *The Socio-Economic Power of Renewable Energy Production Cooperatives in Germany: Results of An Empirical Assessment*. Wuppertal: Wuppertal Institut. http://epub.wupperinst.org/frontdoor/index/index/docId/5364

DECC. (2012). *An Energy Bill to Power Low-Carbon Economic Growth, Protect Consumers and Keep the Lights On*. London: Department of Energy and Climate Change. http://www.decc.gov.uk/en/content/cms/news/pn12_151/pn12_151.aspx

DECC. (2013) *Estimated Impacts of Energy and Climate Change Policies on Energy Prices and Bills*. London: Department of Energy and Climate Change. https://assets.publishing.service.gov.uk/government/uploads/system/uploads/attachment_data/file/172923/130326_-_Price_and_Bill_Impacts_Report_Final.pdf

DECC. (2014a). *Policy Impacts on Prices and Bills*. Based DECCs 2013 Report (op cit), Updated in 2014, Department of Energy and Climate Change, London, Relayed by BEIS. https://www.gov.uk/guidance/policy-impacts-on-prices-and-bills

DECC. (2014b). *An Overview of the Investment Secured in Low Carbon Energy Since 2010*. London: Department of Energy and Climate Change. http://

www.gov.uk/government/publications/an-overview-of-the-investment-secured-in-low-carbon-energy-since-2010

DECC. (2015). *EMR Evaluation*. London: Department of Energy and Climate Change. https://www.gov.uk/government/publications/electricity-market-reform-evaluation

Department of Energy. (1977). *The Prospects for the Generation of Electricity from Wind in the UK*, Energy Paper 21. London: Department of Energy.

Diesendorf, M., & Elliston, B. (2018). The Feasibility of 100% Renewable Electricity Systems: A Response to Critics. *Renewable and Sustainable Energy Reviews, 93*, 318–330. https://www.sciencedirect.com/science/article/pii/S1364032118303897?via%3Dihub

EAC. (2018). *Green Financing Report*. Environmental Audit Select Committee. http:www.parliament.uk/business/committees/committees-a-z/commons-select/environmental-audit-committee/news-parliament-2017/green-finance-report-published-17-19/

Elliott, D. (1995). *The UK Wavepower R&D Programme*. Technology Policy Group Occasional Paper 27, The Open University, Milton Keynes.

Elliott, D. (2015). Green Jobs and the Ethics of Energy. In M. Hersh (Ed.), *Ethical Engineering for International Development and Sustainability*. London: Springer.

Elliott, D. (2017). Renewable Energy in the UK: A Slow Transition. In S. Shmelev (Ed.), *Green Economy Reader : Lectures in Ecological Economics and Sustainability* (Vol. 6). Cham: Springer.

Elliott, D., & Elliott, R. (1976). *The Control of Technology*. London: Wykeham/Taylor and Francis.

EST. (2008). *Power in Numbers*. London: Energy Saving Trust.

Eurostat. (2017). *Share of Energy from Renewable Sources*. Eurostat Data. http://ec.europa.eu/eurostat/documents/2995521/7905983/8-14032017-BP-EN.pdf/af8b4671-fb2a-477b-b7cf-d9a28cb8beea

Eurostat. (2018). *Share of Energy from Renewable Sources*. Eurostat Data. http://appsso.eurostat.ec.europa.eu/nui/show.do?dataset=nrg_ind_335a&lang=en

EWG. (2017). *The Shift from Feed-in-Tariffs to Tenders Is Hindering the Transformation of the Global Energy Supply to Renewable Energies*. Berlin: Energy Watch Group. http://energywatchgroup.org/wp-content/uploads/2018/01/FIT-Tender_Fell_PolicyPaper_EN_final.pdf

Flintoff, J.-P. (2011, May 27). Power to the People: Dale Vince, Green Energy Pioneer. *Daily Telegraph*. https://www.telegraph.co.uk/news/8537788/Power-to-the-people-Dale-Vince-green-energy-pioneer.html

Full Fact. (2013). *How Much Do 'Green Taxes' Add to Energy Bills?* Full Fact Estimates. https://fullfact.org/news/how-much-do-green-taxes-add-energy-bills/

Goodall, C. (2017, March 30). Keeping David MacKay's 'Sustainable Energy – Without the Hot Air' Up-to-Date. *Carbon Commentary*. https://www.carboncommentary.com/blog/2017/3/30/l6qcqgoedse1wmjjz-87t09usoq6jva

Gosden, E. (2015, March 12). Onshore Wind Farm Ban Will Raise Energy Prices, Tory MP Tim Yeo Warns. *Daily Telegraph*. https://www.telegraph.co.uk/news/earth/energy/windpower/11465812/Onshore-wind-farm-ban-will-raise-energy-prices-Tory-MP-Tim-Yeo-warns.html

Greenpeace. (2015). *Energy[R]evolution Report* (5th ed.). Greenpeace/Global Wind Energy Council. http://www.greenpeace.org/international/Global/international/publications/climate/2015/Energy-Revolution-2015-Full.pdf

Harper, P. (2016). Alternative Technology and Social Organisation in an Institutional Setting. *Science as Culture, 25*(3). http://www.tandfonline.com/doi/full/10.1080/09505431.2016.1164406

Heard, B., Brook, B., Wigely, T., & Bradshaw, C. (2017). Burden of Proof: A Comprehensive Review of the Feasibility of 100% Renewable-Electricity Systems. *Renewable and Sustainable Energy Reviews, 76*, 1122–1133. www.sciencedirect.com/science/article/pii/S1364032117304495?via%3Dihu

Heidari-Robinson, S. (2016, July 29th). Personal Reflections on Decarbonization in the UK. *Linkedin Post*. https://www.linkedin.com/pulse/personal-reflections-decarbonization-uk-stephen-heidari-robinson

Heinberg, R. (2017). *Are We Doomed? Let's Have a Conversation*. Corvallis: Post Carbon Institute. http://www.postcarbon.org/are-we-doomed-lets-have-a-conversation/

Hendry, C. (2012). Charles Hendry at the Opening of the Annual All Energy Conference in Aberdeen, as Reported at the Time on the Conference Web Site and in *Renew* 2012, Dec 2012, Oddly This Statement Did Not Appear in the Version of His Speech on the DECC Website.

Hirsh, R. (2011). Historians of Technology in the Real World: Reflections on the Pursuit of Policy-Oriented History. *Technology and Culture, 52*(1), 6–20. https://vtechworks.lib.vt.edu/bitstream/handle/10919/25476/52.1.hirsh.pdf?sequence=1

HM Government. (2015). *UK's Response to European Commission Consultation on Energy Market Design*. HM Government. https://www.gov.uk/government/publications/uks-response-to-european-commission-consultation-on-energy-market-design

Hurd, N. (2017, March 14). *Low Carbon Economy Debate*. House of Commons. https://hansard.parliament.uk/Commons/2017-03-14/debates/7A6CE7A3-A0B0-4E80-A64A-81BD27E77CF8/Low-CarbonEconomy

Ingham, J. (2014, October 15). UK's Wind Farm 'Folly': Electric Bills to Soar by £1000 Thanks to Reliance on Wind Power. *Daily Express*. https://www.express.co.uk/news/uk/522800/Wind-farms-blamed-for-winter-power-cut-and-rise-energy-bills

Jacobson, M., Delucchi, M. Bauer, Z., Goodman, S., Chapman, W., Cameron, M., Bozonnat, C., Chobadi, L., Clonts, H., Enevoldsen, P., Erwin, J., Fobi, S., Goldstrom, O., Hennessy, E., Liu, J., Lo, J., Meyer, C., Morris, S., & Yachanin, A. (2017). 100% Clean and Renewable Wind, Water, and Sunlight All-Sector Energy Roadmaps for 139 Countries of the World. *Joule, 1*(1), 108–121. http://www.sciencedirect.com/science/article/pii/S2542435117300120

Klein, N. (2014). *This Changes Everything: Capitalism vs. the Climate*. New York: Simon & Schuster.

Koirala, P. B, Koliou, E., Friege, J., Hakvoort, R., & Herde, P. (2016). Energetic Communities for Community Energy: A Review of Key Issues and Trends Shaping Integrated Community Energy Systems. *Renewable and Sustainable Energy Reviews, 56*, 722–744. http://www.sciencedirect.com/science/article/pii/S1364032115013477

Lazard. (2017, November 2). *Levelized Cost of Energy 2017*. Lazard International Analysts. https://www.lazard.com/perspective/levelized-cost-of-energy-2017/

Leggett, J. (2018). *Solar Century's History in Pictures and Charts*. https://jeremyleggett.net/2018/05/26/solarcentury-a-history-on-our-20th-birthday-26th-may-2018-in-pictures-and-charts/

Liddell, H. (2000, July–August). *DTI Energy Minister's Speech to a European Solar PV Conference in Glasgow, as Reported in Renew 126*. Renew On-line 27. http://eeru.open.ac.uk/natta/renewonline/rol27/11.html

Lloyd, H. (Ed.). (2018). *A Distributed Energy Future for the UK*. London: Institute for Public Policy Research. https://www.ippr.org/files/2018-09/a-distributed-energy-future-for-the-uk-september18.pdf

Lovegrove, S. (2015, September 5). Quoted in Rees, J., New Power Stations? We'll Just Use Less Electricity. *This Is Money*. http://www.thisismoney.co.uk/money/news/article-3223539/New-power-stations-ll-just-use-electricity-Britain-s-new-Energy-Secretary-outline-plans-Autumn.html

LUT/EWG. (2017). *Global Energy System Based on 100% Renewable Energy–Power Sector*. Berlin: Lappeenranta University of Technology, Finland/Energy Watch Group. http://energywatchgroup.org/wp-content/uploads/2017/11/Full-Study-100-Renewable-Energy-Worldwide-Power-Sector.pdf

MacKay, D. (2008). *Renewable Energy Without the Hot Air*. Cambridge: Self Published. http://www.withouthotair.com/

Martin, A. (2018, February 1). The UK Still Has Some Way to Go to Hit Its 2020 Renewable Energy Target. *Alphr*. http://www.alphr.com/energy/1008375/uk-renewable-energy-progress-2020

Mitchell, C. (2008). *The Political Economy of Sustainable Energy*. Basingstoke: Palgrave.

Monbiot, G. (2010, July 22). How Scrapping the SDC to Save Money Will Cost the Taxpayer a Fortune. *The Guardian*. https://www.theguardian.com/environment/georgemonbiot/2010/jul/22/carbon-emissions-waste

NIC. (2018). *National Infrastructure Assessment, Chapter 2, Low Cost Low Carbon*. London: National Infrastructure Commission. http://www.nic.org.uk/assessment/national-infrastructure-assessment/low-cost-low-carbon/

Paterson, O. (2014, October 16). *Keeping the Lights On*. London: Annual Global Warming Policy Foundation Lecture. http://www.thegwpf.org/2014-annual-gwpf-lecture-owen-paterson-keeping-the-lights-on/

PCS. (2017). *Just Transition and Energy Democracy: A Civil Service Trade Union Perspective*. London: Public and Commercial Service Association. http://www.pcs.org.uk/sites/default/files/site_assets/resources/green_workplaces/2017/Just%20Transition%20&%20Energy%20Democracy%20-%20a%20civil%20service%20trade%20union%20perspective.pdf

Perry, C. (2018, May 25). Parliamentary Statements, House of Commons. *Hansard*. http://www.parliament.uk/business/publications/written-questions-answers-statements/written-question/Commons/2018-05-16/144815/

Pollin, R. (2018). De-Growth v's A Green New Deal. *New Left Review, 112*, 5–24.

REN21. (2018). *Renewables 2018 Global Status Report*. Renewable Energy Policy Network for the 21st Century. http://www.ren21.net/status-of-renewables/global-status-report/

Renews. (2018). *Tory Voters Back Onshore*. Renews Coverage of You Gov Poll. http://renews.biz/88622/tory-voters-back-onshore/Tory voters back onshore

Rudd, A. (2015, June 22). *Parliamentary Statement*. House of Commons, Hansard, Column 617. http://www.publications.parliament.uk/pa/cm201516/cmhansrd/cm150622/debtext/150622-0001.htm#1506227000002

RUK. (2015, September–October). Renewable UK Press Statement as Reported in Renew 217. *Renew On-Line, 117*, 7. https://renewnatta.files.wordpress.com/2014/11/rol117.pdf

Salter, S. (2016). Wave Energy: Nostalgic Ramblings, Future Hopes and Heretical Suggestions. *Journal of Ocean Engineering and Marine Energy, 2*(4).

Science and Technology Committee. (2003). Report of the House of Commons Select Committee on Science and Technology Review of Non Fossil Energy R&D.

Smil, V. (2016). Examining Energy Transitions: A Dozen Insights on Performance. *Energy Research & Social Science, 22*, 194–197. http://www.sciencedirect.com/science/article/pii/S2214629616302006

Smith, A. (2007). Translating Sustainabilities Between Green Niches and Socio-Technical Regimes. *Technology Analysis & Strategic Management, 19*(4), 427–450.

Smith, A., Hargreaves, T., Hielscher, S., Martiskainen, M., & Seyfang, G. (2016). Making the Most of Community Energies: Three Perspectives on Grassroots Innovation. *Environment and Planning A, 48*(2), 407–432.

Sovacool, B. (2016). How Long Will It Take? Conceptualizing the Temporal Dynamics of Energy Transitions. *Energy Research & Social Science, 13*, 202–215. https://www.sciencedirect.com/science/article/pii/S2214629615300827

Spicer, M. (1989, January 30). Parliamentary Answer, House of Commons Debate. *Hansard, 146,* cc 2–4. https://api.parliament.uk/historic-hansard/commons/1989/jan/30/renewable-energy-sources#S6CV0146P0_19890130_HOC_14

Stern, N. (2006). *The Economics of Climate Change*. Report for UK Treasury. Cambridge University Press.

Thurston, C. (2017, November 27). As Mexican Solar Auction Prices Scrape Bottom, Will Quality Be Threatened?. *Renewable Energy World.* https://www.renewableenergyworld.com/articles/2017/11/as-mexican-solar-auction-prices-scrape-bottom-will-quality-be-threatened.html

UCL. (2011, June 29). *Selective Influence: The Policy Impact of House of Commons Select Committees*. Press Release on a University College London Study. http://www.ucl.ac.uk/constitution-unit/constitution-unit-news/290611a

Vaughan, A. (2018, August 22). Green Energy Tariffs 'No Longer an Expensive Luxury', Study Finds. *The Guardian.* https://www.theguardian.com/business/2018/aug/22/green-energy-tariffs-uswitch

Wills, R. (2016, August 31). *Cultural Dimensions of Community Energy*. UKERC Guest Blog, UK Energy Research Centre. http://www.ukerc.ac.uk/network/network-news/guest-blog-cultural-dimensions-of-community-energy.html

Wilson, B. (2002, June 21). *DTI Energy Minister's Comment in a House of Commons Debate on Energy Policy*. https://publications.parliament.uk/pa/cm200102/cmhansrd/vo020620/debtext/20620-15.htm

Wilson, J. C. (2010). *A history of the UK Renewable Energy Programme, 1974–88: Some Social, Political, and Economic Aspects* (PhD Thesis). University of Glasgow. http://theses.gla.ac.uk/3121/1/2010WilsonJohnPhD.pdf

Wood, G., & Dow, S. (2011). What Lessons Have Been Learned in Reforming the Renewables Obligation? An Analysis of Internal and External Failures in UK Renewable Energy Policy. *Energy Policy, 39*(5), 2228–2244. https://www.sciencedirect.com/science/article/pii/S0301421510008360

Woodman, B. (2016, August 12). Hinkley Point C Delay: How to Exploit This Attack of Common Sense in Energy Policy. *Energy Post*. http://www.energy-post.eu/hinkley-point-c-delay-exploit-attack-common-sense-energy-policy

Yang, C.-J. (2016). *Belief-Based Energy Technology Development in the United States: A Comparative Study of Nuclear Power and Synthetic Fuel Policies*. Amherst: Cambria.

Index

D. Elliott, *Renewable Energy in the UK*, Energy, Climate and the Environment,
https://doi.org/10.1007/978-3-030-04765-8

The manufacturer's authorised representative in the EU is Springer
Nature Customer Service Centre GmbH, Europaplatz 3, 69115 Heidelberg,
Germany. If you have any concerns regarding our products, please
contact ProductSafety@springernature.com

Printed and bound by CPI Group (UK) Ltd, Croydon, CR0 4YY
30/09/2026
02225443-0007